Fire Fighting Strategy and Tactics

Featuring an 8-Step Process for Sizeup

by Harry Carter, Ph.D.

Edited By: Lynne Murnane

Published By:
Fire Protection
Publications
Oklahoma State University
©1998

Design/Layout: Ann Moffat
Graphics: Don Burull, Shelley Hollrah, Ann Moffat

ISBN 0-87939-160-x
Library of Congress 98-88706

First Edition

Printed in the United States of America   3 4 5 6 7 8 9 10

Table of Contents

About the Author

Harry R. Carter, Ph.D., is a municipal fire protection consultant with far-reaching fire service experience. Dr. Carter is also an active fire officer who has had extensive career fire fighting experience with the Newark, New Jersey Fire Department. A former Battalion Commander, he currently serves as the Chief of Training and Commander of the Hazardous Materials Response Team. He has also had lengthy volunteer experience. He is a former Fire Chief and Training Officer for the Adelphia Fire Company, Howell Township, NJ. He was the State Fire Marshal for the New Jersey Army National Guard from 1981 to 1991. He has also taught fire risk analysis and community fire defense programs for the National Fire Academy. Harry currently serves as the Second Vice Chair of the International Society of Fire Service Instructors Board of Directors. A long-time member of the National Fire Protection Association, he currently serves on their Fire Instructors Professional Qualifications Committee.

A veteran of over three decades in the fire and emergency service world, he holds degrees in fire service administration, public policy analysis, fire safety administration, the social sciences and business administration. In 1977, Dr. Carter joined the faculty at Ocean County College to develop a new fire science program for them. He served as the first coordinator of that program and worked in that capacity until 1981.

He is the author of four textbooks: *Managing Fire Service Finances, Management Tactics in Fire Protection, Strategic Planning and Fire Protection*, and *Understanding Fire Behavior*, published by the International Society of Fire Service Instructors. He is the co-author of a fifth text, *Management in the Fire Service*, 2nd edition, an NFPA publication.

Dr. Carter has had nearly 600 articles published in all of the major fire service journals and is a contributing editor for *Firehouse* magazine and *The Pennsylvania Fireman*. He writes monthly columns for *The Voice*, and the *1st Responder* news group.

Acknowledgments

I would like to begin my acknowledgments with a simple statement: Writing a book is an arduous and draining process. The only job worse than that of writer is that of editor, who has to make sense of the author's ideas. In my case, I have ended up sounding much smarter than I really am, which is due in no small part to my editor, Lynne Murnane. She appeared to take great joy in shortening my verbiage. She actually thanked me for giving her so many words to cut. Thank you, Lynne.

An author is blessed if he has friends who will tell him how bad his efforts really are, and how they intend to make him right with the world. A tip of the helmet is graciously extended to my friends who took the time to wade through this work:

Alan Brunacini	Phoenix Fire Department
Jack Peltier	(my representative to the world of Common Sense) Firetech Group Inc. of Marlboro, Massachusetts
Sue Peltier	(the lady charged with keeping Jack in line) Massachusetts Firefighting Academy, Stow, Massachusetts
Battalion Chief Ken Folisi	Lisle-Woodridge Fire District, Naperville, Illinois
Captain George Anderson	Newark, New Jersey Fire Department

My very special thanks to Ron Jeffers and Harvey Eisner for their photographic support.

And to my long-suffering wife, Jackie: Dear, I don't know where I would be in life, if you weren't at my side. I lovingly dedicate this work to you.

1

Fire Fighting:
A Historical Perspective

In order to develop an understanding of how fire fighting developed to the point at which it exists today, we must take a look at the perspective of history. The fire service is an extension of the society within which it exists. The forces that shaped the growth and development of America as a nation also guided the evolution of fire protection from its earliest beginnings as a communal effort at self-protection. It was the use of fire to cook our food and warm our homes, along with the ability to grow crops, that began the development of modern civilization. Much of what makes our society a success is in some way related to fire or to the energy created by the combustion process.

As areas increased in development, the need arose for some form of organized protection from the devastation caused by uncontrolled fire. The earliest traceable efforts show that fire protection forces began during the reign of Caesar Augustus in 23 B.C.

A major problem with this force was its source of labor. The slaves who served as firefighters were not especially interested in facing death and danger to save homes that they could never live in. Not surprisingly, they were not particularly effective. It took a serious fire in 6 A.D., apparently another in a long list of similar disasters, to cause the formation of the body of men known as vigiles, whose duty it was to protect the city of Rome. This is apparently the first recorded instance of public funding for fire protection.

A number of important fire protection-related developments came out of England. One of the first recorded fire prevention regulations came during the reign of William the Conqueror. He set down a rule that all cooking fires were to be extinguished each evening and the hearth area covered.

The Great Fire of London in 1666 caused further regulations to be developed. After a large area of the city was destroyed by fire, the city council called for all homes to be rebuilt from brick or stone. These rules also forbade dangerous occupations in the homes of city residents.

Early attempts at colonization in America were almost thwarted by fires that devastated the settlement at Jamestown, Virginia, and that struck at Nieuw Amsterdam, destroying the only ship available at what is now Manhattan Island in New York City. These were but two of the many fires that would race across the face of the infant colony, America. Efforts were fitfully undertaken in various parts of the colonies to deal with fire's potential for destruction.

One of the earliest supporters of fire prevention was Benjamin Franklin. Through his influence, city officials in Philadelphia acted upon his many suggestions for improved fire protection and suppression measures. His writings in *Poor Richard's Almanac* about fire prevention brought important knowledge to the general public. Throughout the colonies, fire's effects were attacked by the development of building codes (prohibiting such practices as thatched roofs), the introduction of no-smoking ordinances, and other similar measures designed to limit a community's fire risk exposure.

Efforts to guard against the financial risks created by fire were stimulated by people such as Benjamin Franklin. In 1752, he founded the first successful American fire insurance company in the colonies. His foresight led to the formation of an industry which, to this day, has been in the forefront of efforts to improve America's response to the impact of fire upon our society. In 1736, Franklin also participated in the formation of a volunteer fire company in Philadelphia.

It is generally felt that Boston was the first city in America to employ full-time career fire fighting personnel at public expense. A captain and 12 other fire personnel were hired in 1752 to provide the manpower, and a pumping engine was purchased from England. Thus did Boston become the first career fire department in America. From an administrative standpoint, it is important to note that the fire personnel in Boston were paid for each run they made and for the training sessions they attended to maintain proficiency. This paid-on-call form of fire department staffing is still quite popular. History records that the first full-time, fully paid career fire department was formed in 1853 in Cincinnati, Ohio. Other major cities followed as our nation grew and prospered during the westward expansion of the 1800s.

Fire fighting equipment has made many strides. In the early decades of the twentieth century, there were just three sizes of hose: 2½-inch, ¾-inch, and 1½-inch. Engine companies were equipped with 2½-inch rubber-lined, cotton-jacketed hose to feed handlines or to siamese into deck guns, monitors, or water towers, and with large rubber suction hoses for taking water from source to pumper. Small, speedy chemical companies used ¾-inch booster hose to deliver their quick burst of attack water. This tradition persisted well into the 1950s with many departments using all 2½-inch hose for attack, whether inside or out. Booster hose was reserved for mop-up operations.

During the 1960s things took a turn for the better with the introduction of 1½-inch attack hose. This was lighter and more maneuverable than any previous attack line. Some fire departments avoided this change until they absolutely had to. This did not make a great deal of sense. It punished people by forcing them to operate with the bulky and awkward 2½-inch hoseline during inside fire attack situations.

On the supply hose front, great strides came from west to east in our country. During the late 1960s, large-diameter supply hose began to come on the scene. With its ability to move large volumes of water over great distances with a minimum of friction loss, it soon gained favor. Like a great rippling wave it moved eastward through the 1970s into the 1980s. Today, 5-inch diameter hose is used in every state in the union. In many instances, the need for pumpers is reduced. A department with large-diameter hose has a reduced need for in-line pumping in areas with good hydrant pressure and spacing. More water can now be moved with less effort. This has had a strong positive effect on fireground operations.

The earliest fire fighting nozzles were the open playpipe nozzles. A great improvement came with the development of the gated shut-off nozzle. Research and development has led firefighters through a wide range of solid stream, solid bore, and fog nozzles. The fire service has now evolved to a point where nozzles are available that can provide a range of flows by increasing the pressure from the pumper supplying the line.

New alloys of metal have reduced the weight of hose couplings, fittings, tools, and portable monitor devices so that firefighters will be suffering fewer back injuries and hernias from working with these devices on the fireground. It is possible now to supply portable monitors directly from large-diameter hose through the fittings and reducers now available.

While we have moved a long way from the use of horses to pull our fire equipment, the basic premises of fire apparatus remain unchanged: vehicles carrying water, ladders, hose, or a combination of these three things along with designated tools, respond to and extinguish fires. The trend in vehicle development has been to move the people off the outside of the vehicle and back into an enclosed cab. Research shows that Detroit was designing this style of fire apparatus back in the 1930s. It took a long time to catch on with the rest of the fire service.

Strangely enough, there is one area where the development of fire equipment has come full circle. Smoke ejectors began to enter the fire service very early in this century. The first large smoke ejector in the United States was developed by the Minneapolis Fire Department. It was truck mounted and appears to have been used successfully for a number of years.

For a time, the technology moved toward the portable units, which I found on entering the fire service in 1966. And now as we move into the twenty-first century, the use of positive ventilation tactics has spawned a new generation of large, truck-mounted smoke ejector fans.

As an Incident Commander, you will need an understanding of the elements that form the equipment base for your fireground operations in the years to come. Here is a list of fire service equipment items to stimulate your thinking. You will need to know about each if you are to become knowledgeable in the use of fire fighting strategy and tactics.

- Larger, more powerful pumpers with enclosed crew cabs
- Sturdier aerial and ground ladders
- Later generation tower ladders and snorkel equipment
- Large, mobile air resupply systems
- Large, mobile smoke ejection and ventilation units
- Personal alert safety devices (PASS) to let us know when our firefighters are in trouble
- Stronger ground ladders
- Better SCBA seat-mounted brackets
- Flowmeters that show the flow of water, rather than the pounds per square inch
- Large-diameter hose and many special fittings to enhance its use and flexibility
- Improved hydrant security devices and wrenches for high-crime urban areas

- New portable pumps that allow you to achieve usable water flows from rural water sources

To this day, the majority of fire departments in our country maintain the proud traditions begun by the volunteers of colonial America. More than a million volunteer firefighters staff over 21,000 fire departments. Contrasted with this are figures indicating that approximately 185,000 career firefighters staff approximately 1,600 fully paid fire departments.

2
Understanding Fire Behavior: Know Your Enemy

A knowledge of fire behavior is essential to the development of one's skills as a firefighter. Experience has shown that for fire professionals, a knowledge of fire behavior is critical. After all, how can a fire be attacked properly if you do not know what fire is, how it behaves, and how it travels? You must be able to recognize potential fuel packages in a building or compartment, and use this information to estimate the fire growth potential for that space or the building as a whole.

This forms an important part of your assessment of the fire. This can be part of your sizeup questions:

- Where is the fire?
- Where is it going?
- Where have I got to stop it?

(**NOTE:** These sizeup questions will be covered in more depth in Chapter 8.)

The goal of this chapter is to assist you in developing an understanding of fire behavior. The development and spread of fire depends on certain physical principles. We will present a discussion of the essential concepts from chemistry and physics. They are the basic building blocks in the development of a fire attack. The Incident Commander must understand how fire works and how to control it by manipulating its physical and chemical principles.

COMBUSTION

Fire and combustion are terms that are often used interchangeably. Technically, however, fire is a form of combustion. Combustion is a self-sustaining chemical reaction yielding energy or products that cause further reactions of the same kind. Fire is a rapid, self-sustaining oxidation process accompanied by the evolution of heat and light of varying intensities. The speed of a fire reaction ranges from a slow, smoldering fire, to the rapid fire of an explosion (Figure 2.1).

2.1 As we all know, when the combustion process results in fire, the result is a lot of heat, light, and danger. *Courtesy of Ron Jeffers.*

Further, oxidation produces energy. In a fire, the resulting energy takes the form of heat and/or light. The speed of the reaction affects the way in which the energy that is released appears to those who see it. The speed of the reaction determines whether the observer feels a lot of heat and light (fast) or very little heat (slow). Two examples of slow oxidation would be the rusting of metal and the yellowing of newsprint. An example of rapid oxidation would be the combination of oxygen and gasoline in the presence of a heat source (Figure 2.2). The standard symbolic depiction of the components of fire is in the form of a tetrahedron (Figure 2.3).

The graphic depiction of the elements of fire is an excellent starting point to discuss fire extinction. It is simple, really: All you have to do is eliminate any part of the tetrahedron: remove the fuel, reduce the heat, exclude the oxygen, or interrupt the chemical chain reaction. Using water to cool a fire, or dirt or sand to smother a woods fire are but two of the many examples one could choose. The interruption of the chemical chain reaction of the fire is yet another way that fire personnel can extinguish a fire. It is up to the Incident Commander to use those methods available within the fire fighting arsenal. It is important to deploy the appropriate method at the right moment.

When the four components of the fire tetrahedron come together, ignition occurs. For a fire to grow beyond the first material ignited, heat must be transmitted beyond the first material to additional fuel packages. When a fire occurs outside, the heat rises up and moves away. In a closed space, however, heat builds up and fire can spread, depending on the configuration of the space and any openings that may exist.

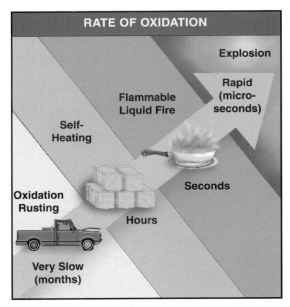

2.2 Combustion, a self-sustaining chemical reaction, may be very slow or very rapid.

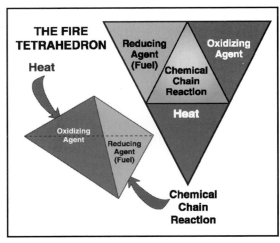

2.3 All components of the fire tetrahedron must be present for fire to occur.

WHAT IS BURNING?

It is well known that the products of combustion can cause death and destruction. These products are determined by the material that is burning or may come to be burned. As the concept of fuel is applied to oxidation, a fuel is any material that can be oxidized. This resulting oxidation produces heat and light.

It is important to remember that some fuels present a greater danger than others. Consider the fire involving flammable liquids. Then compare that to a wood-fueled blaze. The flammable liquid gives off more heat. The Incident Commander has to consider the nature of fuels involved as part of the "What have I got?" question. An early error caused by missing this clue can lead to later disaster.

FIRE DEVELOPMENT

The Incident Commander must know that there are five stages of fire (Figure 2.4). Intervention may occur during any one of them. They are:

- Ignition
- Growth
- Flashover
- Fully developed
- Decay

As stated, ignition occurs when the four components of the tetrahedron come together. Shortly after ignition, a fire plume begins to form above the burning fuel. As temperatures rise, heat buildup occurs throughout the compartment. The fire will continue to grow as long as sufficient fuel and oxygen exist.

Flashover is the transition phase between the growth and fully developed stages. During this phase, radiant heat causes pyrolysis in the combustible solid materials in the compartment. At some point, the compartment flashes over, and the fire reaches the fully developed phase.

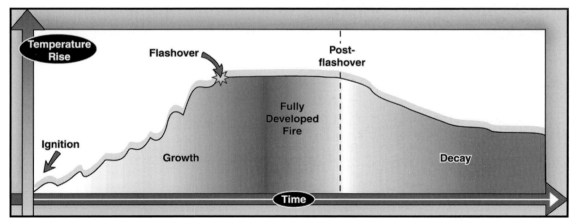

2.4 Fire follows predictable stages of development.

As the fire consumes the available fuel in the compartment, the rate of heat release begins to decline. At this time there are high temperatures and glowing embers.

FACTORS THAT AFFECT FIRE DEVELOPMENT

As the fire progresses from ignition to decay, several factors affect its behavior and development within the compartment:

- Size, number, and arrangement of ventilation openings
- Volume of the compartment
- Thermal properties of the compartment enclosures
- Ceiling height of the compartment
- Size, composition, and location of the fuel package that is first ignited
- Availability and locations of additional fuel packages (target fuels)

FIRE EXTENSION

If a fire can be contained to the compartment of origin, extinguishment is a relatively simple matter. However, a great many problems come from the extension of the fire to locations outside the compartment of origin. Fire can extend in the following ways:

- Through unprotected openings
- Through multi-layered ceilings
- Laterally in cockloft spaces
- Upward through openings in walls
- Rising upward in high-rise situations (Figure 2.5)
- Thermal layering

These facts form a basis for your fireground decision-making process.

ROLLOVER AND BACKDRAFT

Flameover and rollover are two terms that describe a condition where flames move through or across the unburned gases during a fire's progression. Flameover is distinguished from flashover by its involvement of only the fire gases; rollover involves the surfaces of other fuel packages.

My fellow firefighters and I have been driven from many a fire building by a wave of flames rolling across the ceiling toward us (Figure 2.6). Research indicates

2.5 This fire may well extend to the floors above. *Courtesy of Harvey Eisner.*

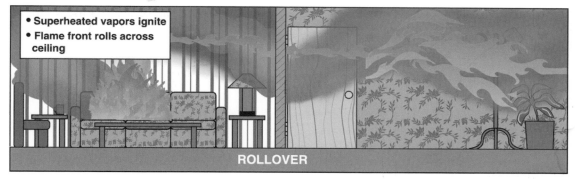

- Superheated vapors ignite
- Flame front rolls across ceiling

ROLLOVER

2.6 As the hot gas layer forms at the ceiling of the compartment, the flame front extends across the ceiling.

that a layering of vapors occurs, which is also known as heat stratification, and thermal balance. We use thermal layering as part of our operation when we vent over a fire compartment.

A greater danger comes from the creation of backdraft conditions, with their potential for explosive release. As a fire grows, large volumes of unburned gases can accumulate in unventilated areas. Any action that allows air to mix with these gases has the potential to create a backdraft explosion. The following conditions may indicate the potential for a backdraft:

- Pressurized smoke exiting small openings
- Black smoke becoming dense grey yellow in color
- Confinement and excessive heat
- Little or no visible flame
- Smoke leaving the building in puffs or at intervals (appearance of breathing)
- Smoke-stained windows

2.7 Being able to recognize the potential for backdraft and taking steps to eliminate it are some of the most important safety decisions an Incident Commander can make.

By recognizing these indicators, precautions can be taken that can lower the risk. Proper ventilation can reduce the danger from backdraft (Figure 2.7). These early clues can lead to some of the most important decisions that an Incident Commander can make. A properly vented structure is easier to enter. A fire attack can proceed more easily under these circumstances.

It is important to remember that there have been some significant changes in fire growth characteristics over the last few decades. They have come as a result of increased fuel loads from the use of synthetic building materials, interior finishes, and furnishings. In addition to the much larger fire loads, many of the new materials have greater burning rates than those used years ago.

The introduction of these new types of combustibles in North America brought about new challenges for the firefighter, in that we are often seeing more fully developed fires upon arrival. As this information clearly indicates, changes in fire behavior have created a more dangerous environment for firefighters and this trend will likely continue. An in-depth knowledge of fire behavior is essential for a safe and effective fire fighting effort.

SUMMARY

A knowledge of fire behavior is a critical part of the Incident Commander's bag of operational tricks. By understanding what a fire is, how it acts and reacts, and how it travels, Incident Commanders can deploy their resources in the safest possible way. Fire is the enemy and you must learn all you can about it.

3

Fire Fighting Strategy and Tactics: An Introduction

It has been said that we in the fire service perform at our best when surrounded by fire, smoke, and flames. All of our resources are brought together to control the enemy: fire. It might appear to the unknowing observer that fire departments indulge in a frenzied, uncoordinated binge of activity on the fireground. We who do the work know better. The well-trained fire department performs a number of highly interrelated and interconnected activities according to set guidelines. Performing certain of these tasks in an incomplete or out-of-order sequence may expose the offenders to the death penalty.

Fire suppression operations involve the use of a fire department's resources to combat a fire. It has been said that operational success in a fire fighting operation depends upon the ability of a fire department to effectively and efficiently use the available resources to protect lives and property. To ensure that a single coordinated effort is carried out during a fire fighting operation, command authority is vested in a single Incident Commander. This individual is responsible for the development and use of the appropriate **strategy and tactics** used to combat a fire.

According to the *NFPA Fire Protection Handbook*, 18th ed., "Strategy involves the development of a basic plan to deal with a situation most effectively. The plan must identify major goals and prioritize objectives. Strategic decisions are based upon an evaluation of the situation, the risk potential, and the available resources." It is the task of this text to spell out what an Incident Commander needs to accomplish through his/her operating personnel. Basically, this involves answering a number of simple questions:

- Who is needed to perform the operations?

- What are they supposed to do?

- Where on the fireground will these tasks be performed?

Tactics are the actual methods of operation employed by individuals or groups of companies to achieve the strategic goals identified by the Incident Commander. The tactics are employed to answer additional questions that complement the three listed above:

- Who is needed to do what where on the fireground? In what way?

The Incident Commander has two basic options: pursue an offensive attack, which would involve an aggressive attack on the fire at its heart, inside the fire building, or pursue a defensive fire attack. This is where forces are massed on the outside of the building to deliver an enormous amount of water from large-caliber devices such as deck guns, ladder pipes, and 2½-inch handlines with playpipes attached. The key to this selection lies with the Incident Commander's answer to the following simple questions:

- How much fire department attack force do I have?
- How much fire must these forces attack?
- Are there lives endangered that can be safely rescued?

Where there is more fire than attack force, the choice would probably be defensive in nature. Where there are sufficient forces to mount an interior attack, and conditions are not too severe, then the offensive option would be selected. Adding endangered occupants complicates the decision. The *NFPA Handbook*, 18th ed. states that "...rescue is the only acceptable reason for exposing firefighters to unnecessary risks." This is an extremely critical decision on the part of the Incident Commander

To help you understand the organization that an Incident Commander must use to combat a hostile fire, let us look at the following basic generic diagram of how resources might be organized (Figure 3.1):

Each level of the organizational chart is charged with the conduct of a particular aspect of fireground operations. Engine companies, truck companies, and rescue companies are deployed based upon the demands of the specific fire at hand. As you might imagine, the responsibility for developing strategy rests with the Incident Commander. The various sector commanders work to implement the tactical functions necessary to meet the strategy outlined by the Incident Commander. Each individual fire fighting unit works to accomplish tasks that fall under the tactical functions.

Just what are the tactical functions fire fighting units must perform at the scene of a fire? Classically, they have been listed as:

1. Rescue endangered lives.
2. Protect exposures.
3. Confine the fire to the smallest area possible (room, floor, building, block, or city of origin).
4. Extinguish the fire.
5. Ventilate to save lives and make a safer working environment for the firefighters.
6. Overhaul to fully extinguish fire and find the point of origin.
7. Conduct salvage operations to limit damage to the fire building and contents.

Let us look at a few of the key points. My experience has been that tactical operations cannot always follow along in precise, textbook fashion. The only given is to place **rescue** at the top of your list of things to do. People exposed to smoke, heat, and flames have but a short time to live. You must take all **REASONABLE** risks to save those lives.

However, using firefighters in situations that are obviously too dangerous should be avoided at all costs. **FIREFIGHTERS' LIVES SHOULD NEVER BE RISKED SAVING PROPERTY**. You must work to ensure that your personnel operate as safely as possible. This can be best accomplished by thorough training and a solid set of General Operating Guidelines (GOGS). We shall discuss these GOGS in a later chapter.

Exposure protection is placed ahead of extinguishment in a number of sources. However, I have found that in many instances, an aggressive interior attack on the seat of the fire can do more to limit the spread of the fire than several master streams later in the operation. If the risk is REASONABLE, take it.

Confinement is often a matter of degree. If you have a one-room fire, your task is to confine it to that room. If you have a much larger fire, say an entire building or a warehouse, your task necessarily changes accordingly (Figure 3.2). I can remember a fire where the deputy chief's strategy was on a much grander scale. Because of the fire's size, his decision to limit the fire to the block of origin seemed prudent and reasonable.

Extinguishment is also a matter of degree. An old bit of wisdom states that the majority of fires are extinguished with a single line. It should be your objective to see that the correct type and number of hoselines are placed to extinguish the fire. A failure to "stretch heavy" early in a fire fighting operation can doom it to failure. My personal maxim for this decision is simple:

"Big Fire = Big Water; Little Fire = Little Water."

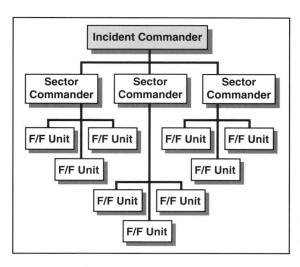

3.1 It is essential that the Incident Commander understand basic guidelines for commitment of resources.

3.2 When you are facing a large fire, your plan for confinement must be adjusted accordingly. You may do well to keep it to the building of origin or the block of origin. *Courtesy of Harvey Eisner.*

This is a simple, easy-to-remember starting point for your fireground operation. Remember, it is often the simple things that trip us up on the fireground.

Inadequate ventilation gives this author nightmares. As one who has spent more than a decade as an engine company commander, my opinion is firm. The work of attack hoselines is made easier by companies that have practiced effective ventilation techniques. This feeling has permeated training in a number of fire departments. Remember, you can save lives as well as extinguish a fire more quickly and effectively by opening up a fire building.

Anyone who has ever had a rekindle is aware of the importance of complete and thorough overhaul of fire-damaged buildings. The effective Incident Commander will ensure that overhaul begins as soon as possible after extinguishment and proceeds in a systematic manner from the damaged to the undamaged parts of the fire building. Be on the lookout for clues that a suspicious fire has occurred, and preserve any evidence found.

Salvage in many areas has almost become a lost art. The protection of personal property and removal of excess water perform a distinct service to the public you are sworn to protect. Excessive damage has occurred in many cases because firefighters were untrained in salvage skills or just plain lazy. Do not fall victim to this evil.

SUMMARY

In this chapter, we have discussed the two main components of fireground operations: strategy and tactics. They are skills that must be learned by outside study. After the knowledge has been gained, it must be practiced during frequent live drills, chalkboard sessions, and further reading. And seeking knowledge and training must continue throughout your career. Fire fighting strategy and tactics are truly an example of that old adage, which states: "Practice makes perfect."

4
Hoseline Operations

Firefighters are in almost universal agreement that their primary function at the scene of a fire is the saving of lives. Next in importance would be the concept of property conservation, without endangering personnel. And lastly, we must all be concerned with the impact of fire losses on the community in general. This would involve the loss of people and property.

The primary mechanism for the control of fire, and therefore the prevention of death and injury, involves the prompt application of extinguishing agent on the base of the fire. This can best be accomplished with the proper placement of hoselines. In those cases where the volume of fire hinders an aggressive interior attack, the use of master stream appliances is called for.

Let me offer a bit of advice that I have committed to memory from one of the first fire fighting texts I ever added to my library. It has served me well for many years. In his outstanding text, *Fireground Tactics*, the late Emanuel Fried noted, " ... it is important to note that the judicious use of a hose stream ... may save more lives than raising ladders to rescue persons visible at the windows." Little did I know when I read this statement more than two decades ago, just how true it would be.

I have spent a great deal of my career as a firefighter and an officer on engine companies. On many occasions, one hoseline was able to hold a critical stairway position. In doing this, we provided an unbroken means of access for those civilians trapped above the fire floor. And we were able to do this with a smaller number of people. Remember, rescue via a ladder usually involves a one-on-one situation. One firefighter is completely involved in saving one person. Much better to have an open route to safety being protected by an engine company hoseline.

It is not my intention in this book to tell you how to lay a hoseline. There are a number of excellent IFSTA publications for identifying this basic information. Rather, I would like to emphasize some principles for the general use of those skills that are best developed in basic drill evolutions. I call them **Carter's Hoseline Hints**. In both strategic and tactical situations, you will be required to blend a variety of hoseline advancement options in order to achieve your overall goals. Failure to keep an eye on how your hoselines are working can spell doom for your fireground operation.

The first hint is a very basic one. More than 20 years ago, my first pump operations mentor in the Newark Fire Department gave me my **Engine Company Rule #1** for the driver. Your job is to find the water! And you must be sure to get it to the people in your company. Once they are supplied, help as many other people as possible. That guy could find water no matter where we went. And the lesson stuck. All of my people know that water is the key to success.

In the larger urban and suburban areas, you must be sure to select the best hydrants for the task. This is done during your pre-incident planning operations. In rural areas, drafting sites must be identified ahead of time. Knowing where the drafting sites are located in your community can save a great deal of time, when time really counts. So **Hoseline Hint #1 is, be sure you have a source with enough water for your needs** (Figures 4.1 and 4.2).

Hoseline Hint #2: Use the proper size of supply hoseline to get the most from your ample source of water. With the advent of large diameter hose, there is no excuse for failure to move enough water (Figure 4.3).

A simple common sense statement forms the basis for **Hoseline Hint #3. Do not shoot water at smoke**. On countless occasions during the past three decades, I have seen this basic concept violated time and again. Shooting water at smoke causes the wet smoke to lose its heat-related upward lift and drop to the floor of the fire area. The poor visibility will make your job a great deal more difficult. Your search and rescue efforts will be slowed, and more important, the smoke may trap building occupants who are attempting to exit the building under their own power. They may slow your operation, block your progress, and require one or more firefighters to remove them from the burning structure. Smoke

4.1 The engine company's most important job is to secure an ample water supply. *Courtesy of Harvey Eisner.*

4.2 During an emergency is a poor time to search for drafting sites. Use pre-incident planning to identify them ahead of time.

4.3 Maximize the effectiveness of your water supply by using large-diameter supply hose. *Courtesy of Harvey Eisner.*

will also slow your progress as you move your attack lines in toward the fire. So while it may be a difficult, hot, and dirty journey, push your lines in to the base of the fire before you open up with the water. It is best, whenever possible, to move in toward the seat of the fire and apply your water there.

(**NOTE:** The only exception to this rule is when your senses tell you that the potential exists for the possibility of a flashover/rollover as you move in to attack the fire. A quick burst of water above can delay this possibility until proper ventilation is available. However, you must be very careful under these circumstances. If you are unable to quickly attack the fire, you may wish to pursue the attack from a safer location.)

Hoseline Hint #4 is directly tied to the last rule. **Your greatest chance of success lies with making an aggressive interior attack on the seat of a fire, whenever possible.**

The problem with this rule is that people tend to apply it when it is too dangerous to enter the fire building. We tend to see too much of this **Kamikaze fire fighting**. You should know about this type of operation. Think of the last time you saw a crew of firefighters going off to attack a barn fire or a warehouse blaze with a booster line.

Hoseline Hint #5 is very simple. **BIG FIRE = BIG WATER; LITTLE FIRE = LITTLE WATER.** Like beauty, big is in the eye of the beholder. What a firefighter in a small department might consider as big could be quite different from what a firefighter in a large urban area would consider big. Figures 4.4 and 4.5 are two examples of what I am talking about:

4.4 A big fire will require a lot of water and resources to bring under control. *Courtesy of Ron Jeffers.*

4.5 Little fire = little water (but still enough to do the job effectively). *Courtesy of Harvey Eisner.*

You will become better at applying this rule as you gain experience. You must first come to understand just how much fire a particular size hoseline will extinguish. You will develop your confidence about what size hose to use under what situations through a process of study, drill, trial and elimination. But never lose sight of the basic rule of **Big Fire = Big Water: Little Fire = Little Water**. It will serve you well.

Hoseline Hint #6 should be a matter of common sense, but it is often overlooked. **When moving a line up a stairway, close the nozzle**. How do you do this? You extinguish as much fire as you can reach from the base of the staircase. Then shut the nozzle and move up the staircase. As soon as you are set, spot the next area of fire and attack what you can reach. Moving any attack hoseline, regardless of size, is a difficult proposition anyway. By shutting the nozzle, life will become just a bit easier. If you doubt the wisdom of this hint, try it both ways. I have found this method to be much easier.

Hoseline Hint #7 can save you from a great deal of harm. **Never position yourself directly in front of a door to a burning room**. If you are standing upright when the door is opened and it flashes or rolls over, you will be burned. Stay low and be ready.

As a veteran of many decades in the fire business, I could tell you countless stories about being caught in a crowd while attempting to move into or out of a fire area. Therefore **Hoseline Hint #8** can become a matter of self preservation. **Do not crowd the attack crew** (Figure 4.6).

Like any other element of a labor-intensive operation, the people on the tip need a certain amount of room to maneuver. They also need a way out if things go bad during the attack operation. Think of it as the Anvil Chorus: listen for the sound of the clanging air pack tanks. It's a bit harder now with the lightweight cylinders. The clicking is not as loud as the clanging. However, the concept is the same. It is the duty of the interior sector commander to ensure that resources are being deployed in a safe manner. Crowding up is neither a safe nor wise use of resources. So be sure to deploy your people wisely.

The next recommendation is just plain common sense, but sometimes I think there is just not as much common sense running around loose as there once was. **Hoseline Hint # 9** tells us to **be sure that we never pass a fire**. It may be necessary to have the attack team include personnel with forcible entry tools. In this way, the coordinated effort can improve their chances of not passing a fire. They will be able to open up above, below, and on the sides of the attack avenue. Many times we have heard or read about people being trapped on an upper floor by a fire that was bypassed on a lower floor. It has been my experience that only in the rarest of circumstances has passing a fire ever been warranted. As you deploy your attack forces, you should handle fire as you encounter it.

How much safer will it be if your crew on the first floor is handling that area of fire, while your next hoseline crew moves by them to move up the staircase? In this way, you are able to operate in a progressive manner, with each line protecting the next. This allows you to deploy your forces in depth. And it is usually much safer.

I have come to regard **Hoseline Hint #10** as a rule that should never be violated. **An engine company must work as a team**. They must combine their efforts for one purpose: the delivery of an effective fire attack stream. That stream can range all across the board. It might be a booster line being used to snuff out a garbage can. Or it can range up to a mighty master stream. Whatever the mission, the engine company must act as a team. Never forget this. Freelancing by engine company members can be fatal.

Hoseline Hint #11 is extremely simple, but can save a life. Regardless of which type of personnel accountability system your fire department uses, **keep a list of your people in your pocket**. How hard is it to keep a card with your members listed on it? I have kept such a list for a long time. It dates back to a fire many years ago when I was a brand new company officer. We had a major collapse and found it very difficult to account for everyone. To this day, I carry a copy of our Battalion Roster with us. Fortunately there has never been a need to ensure that all of our people were safe. But I am ready (Figure 4.7).

Very few fires have ever been extinguished without the use of some combination of pumper apparatus, water, and hoselines being used by well- trained firefighters. If you are to succeed in the fire fighting business, you need to know how to effectively deploy your engine company people. To assist you toward that end, let us sum up with **Carter's Hoseline Hints:**

4.6 See that the attack crew has enough room to maneuver safely. *Courtesy of Mike Wieder.*

4.7 Make sure you can account for everyone on the scene. *Courtesy of Ron Jeffers.*

Hint #1: Be sure you have a source with enough water for your needs.

Hint #2: Use the proper size of supply hoseline to get the most from your ample source of water.

Hint #3: Do not shoot water at smoke.

Hint #4: Your greatest chance of success lies with making an aggressive interior attack on the seat of a fire, whenever possible. Be aggressive, it can save lives.

Hint #5: BIG FIRE = BIG WATER: LITTLE FIRE = LITTLE WATER.

Hint #6: When moving a line up a stairway, close the nozzle.

Hint #7: Never position yourself directly in front of a door to a burning room.

Hint #8: Do not crowd the attack crew.

Hint #9: Never pass a fire.

Hint #10: An engine company must work as a team to perform as single function. No freelancing.

Hint #11: Keep a list of your people in your pocket.

And once you have committed all of these hints to memory, drill on them. They are the basics and they can save your life.

5

Engine Company Operations

The last chapter offered some basic hints on how to deploy hoselines in a fire fighting scenario. This chapter will discuss the use of engine companies in a fire fighting situation. There is one abiding phrase that you must commit to memory if you are ever to succeed in the delivery of engine company services. Simply stated:

"Water is the name of the game."

No other basic reason exists for the fire pumper as we know it today. While I am sure that there are a great many services that your engine companies provide, anything other than water delivery is just frosting on the cake. In many fire departments, people lose sight of this fact. As my first mentor once taught me, "Kid, your job is to get water for your company. Whether you have to lay in, stretch back, or jump through hoops, get water to the tip." I have done this for many years. And one other important tip that I have learned the hard way goes something like this:

"Big Fire = Big Water; Little Fire = Little Water"

Many fires have been lost because of the improper deployment of engine company resources in a tactical situation. This section will address the following topics:

- Operational mode
- Water source selection
- Hoseline selection
- When to use master streams
- Hoseline placement
- Hoseline advancement

DETERMINING THE ATTACK

Much has been written about the deployment of resources upon arrival at a fire incident. Perhaps the element that provides the greatest direction in how you will attack a fire comes from the mode of operation. Are you going to enter the building and go one-on-one with the fire? Or are you going to remain outside and limit the fire to the block of origin? This is a critical decision. And a great deal of the information that you will use to make this choice comes from your day-to-day contact with your response district. Occupancy types will govern attack decisions almost as much as location and size of any fire you might face.

Would you be as willing to commit troops to an 0300 hours fire in an explosives warehouse, as you would a suburban tract home? I sure hope not! The use of firefighters solely for property conservation is never a good choice. You would be hard pressed to defend such a use of personnel. On the other hand, the judicious use of an aggressive interior attack is justified in a residential scenario where human life is at stake (Figures 5.1 and 5.2). Just be sure to note the word judicious.

Just remember one important fact. As the Incident Commander you are ultimately responsible for the safe deployment of your human resources. It will be your education, training, and experience that will guide you in weighing risk against benefit, and then choosing whether to fight a fire from inside or outside any given structure.

TYPES OF ATTACK

I have found that there are four different operational methods that you will have to choose from:

- Aggressive interior attack (The Cavalry Charge)
- The blitz attack and move in method
- The blitz attack and think about moving in method
- The "keep to the block of origin" operation

What the first-due unit sees as they roll in to the fire scene will determine which of these attack options you will use. But remember, be flexible. The deployment flow between the various attack options must be fluid. If you see that your first choice is not working, be prepared to use another. Factors beyond your control can affect how the fire fighting operation will proceed. Such factors as construction, occupancy, wind, are aided or hindered by the skill and quality of your operational personnel. A fairly routine fire can be lost due to a weak fire attack. And a difficult fire can be tamed by a determined effort.

5.1 If situations permit, an interior attack to save human lives is appropriate. *Courtesy of Harvey Eisner.*

5.2 When it is obvious that a fire is too large or too advanced for firefighters have a chance to rescue someone, a more cautious fire fighting approach is indicated. *Courtesy of Harvey Eisner.*

Determining Resources

At this point I would like to give you a short equation to describe the fire vs. fire fighting effort:

$$\frac{\textbf{How much fire do I have?}}{\textbf{How many resources do I have to throw at the fire?}}$$

If you are blessed with a vast array of resources and are facing a half-vast fire, you should be able to muscle the fire under control in fairly short order. However, it is during those close calls where you and the fire are fairly evenly matched that things will be a bit touchy. This is a corollary to our earlier advice, which stated **Big Fire = Big Water; Little Fire = Little Water.**

Here is a quick example: What should you do if you are facing a fully involved two-story development or sub-division home and you know that you only have two poorly staffed engine companies rolling? Your options will be limited. It is my opinion that you will need at least *9* people responding on two pumpers to have a fighting chance. This group will deploy a minimum force of three 1¾-inch or 2-inch attack lines (Do not forget the two-in two-out rule), while maintaining pumper operators and a safety person, in addition to you as the Incident Commander. Remember that staffing for truck company operations is in addition to the force mentioned above. You are probably looking for a force of 12 to 15 firefighters to accomplish all that you need to do.

You might get away with two lines, but you should always approach fireground strategy and tactics as a realist. How much fire can you kill with tank water? Where will your water supply come from? And how much time will it take to secure it?

Interior Attack

Let us now describe each of the attack styles mentioned above. You should consider an aggressive interior attack in any situation where life is at risk. I have been a part of many fireground operations where a good quick attack on the seat of a fire has saved lives. Factors that you must consider in the choice of an interior attack are:

- Is there life at risk?

- Are there enough firefighters?

- Is there sufficient on-board tank water to make an immediate impact on the fire?

- What is our source of on-going water and can we reach it quickly?

- Is it just plain stupid to send my people into the burning building I see in front of my eyes?

Items 1 and 5 are the true limiting factors in deciding when and where to move attack lines into a burning building. Don't discount gut instincts. If it looks like a dumb thing to do, then you probably should not do it. And if you do not have enough people to run the necessary attack and supply hoselines, then do what you can until help arrives.

I can remember riding in short-handed on a tenement fire one evening. As the driver put the pump in gear and the side man took a supply line to the hydrant, I moved the 1¾-inch attack hoseline back and forth from window to window on the front porch. I managed to knock down a lot of fire before the next due company moved in with me to attack the balance of the fire. Please note that I **DID NOT** enter the building alone. That would have been really dumb.

Where your fire department is located can also have a tremendous bearing on when and how you conduct inside fire fighting operations. It is one thing to operate in a community where there is a reliable water system and adequate hydrant spacing. It is quite another when you must depend on drafting positions, tanker shuttles, and hoseline relays. The certainty of water is a critical component in deciding to run an attack line into a burning structure.

Your level of training and practical experience are also important concerns. If you drill frequently in structural live fire scenarios, you will be more comfortable moving into a burning building. And if you attend a great many such fires in any given year, you will be better prepared to do your thing. You will gain confidence as you gain experience (Figure 5.3).

Blitz Attack

The blitz attack can bring a seemingly difficult fire under control in short order. To do it successfully, however, you need personnel, equipment, coordination, and practice. To accomplish this style of fire fighting, you will need an attack pumper with a water tank capacity of at least 750 gallons and a pre-piped master stream located for ready use (Figure 5.4). The key to success if you are going to use a blitz attack is the quick accomplishment of simultaneous tasks:

- Put the master stream into play
- Stretch a quick supply line
- Stretch a quick attack line
- Do not enter the building until the fire is knocked down and your crew is reassembled.

Once the fire is darkened a bit and your crew is back together, you can then proceed with an interior attack. It will take training and drill ground practice to become proficient in this operation.

5.4 Remember, Big Fire = Big Water. A master stream will be needed to put a lot of water where it will do the most good. *Courtesy of Harvey Eisner.*

5.3 A well-trained attack team will be prepared to face the real thing. *Courtesy of Ron Jeffers.*

But the results can be well worth the effort. I can remember cutting off a fire in a laundry warehouse with just this type of multi-pronged effort. It is another fine tool for your fire fighting arsenal.

When it comes to a blitz attack and thinking about moving in, my best advice is to practice a lot and use this when your brain says, "maybe, maybe not." While your troops are deploying the big water to whack the big fire, your primary responsibility lies in the world of "What if?" Some of the good what ifs are listed here for your review:

- What if we do nothing but hit it with big water?
- What if we move a line in?
- What are the chances of **safe success?**

Safe Success is a concept you will need to rivet into your brain. Success can have a price. Is the fire you are facing worth the added price you may pay for a quick trip into the world of smoke and flames? The what ifs here can have a tremendous downside. You should only move in where there is a good reason. Some of these reasons are:

- You discover a rescue problem.
- The building appears sturdy enough.
- The fire becomes small enough.

Defensive Attack

If you choose to keep hitting the fire with big water, you have immediately moved into the next phase. This is the "keep it to the block of origin" operation. The line between these two phases can be blurred occasionally. Always err on the side of safety (Figure 5.5). What sense does it make to let a building burn for two hours and then put people into it for overhaul operations? On a number of occasions we have just kept water flowing until there was no more smoke. This is not very scientific, but it is safe.

5.5 If the fire is big enough, the defensive attack is the prudent way to go. *Courtesy of Ron Jeffers.*

WATER SUPPLY

Once you have determined the attack mode for the fire you face, you are immediately confronted with another problem. Where will the water for the attack come from? A dark and stormy night is not the time to think your first water supply thought.

In any analysis of municipal fire protection, it is critical to remember the importance of water supply and its delivery to the fireground. Despite the many technical advances in the field of fire protection, water remains the primary extinguishing agent of choice used by fire departments to attack and control fire. The relative ability of a fire department to take the existing water supply and deliver it through their pumper apparatus is perhaps the most critical measure of a fireground operation. Whether it is the direct application of water through hoselines or the support of fixed fire protection devices in a given structure, a fire department's success or failure depends on the manner in which this essential task is performed.

Let us pause and consider the basic components of a municipal water supply system, which are many and varied. Basically, the system can be described by the following fundamental components:

- The source of supply

- The processing or treatment facilities

- The mechanical or other means of moving water

- The distribution system, including storage

While each of these components is important to the successful operation of a municipal water supply system, we will limit our comments to those segments that are essential to fire protection, namely the source of supply and the distribution system.

We depend upon water to extinguish fire by absorbing the heat produced by the combustion process. It does this best when delivered in a spray form, early in the onset of the fire scenario. That, in a nutshell, is the job of the engine company. What, then, are the choices you can make for water supply?

- Hydrants (Figure 5.6)
- Drafting points (Figure 5.7)
- Water tankers
- Portable tanks
- Portable pumps

It is best to drill on the use of each of these sources during the course of a year. While it may be that the bulk of your fire fighting water comes from hydrants, be sure that your people can operate from a draft or with tanker support. Recently I attended my first fire where portable water source tanks were placed into service. While this might be a common event in your fire department, it is not used in either of the departments with which I am very familiar. Photos were taken and blended into my personal water supply training program, just in case. You need to be well-trained and flexible in the use of any water source that might suddenly become a necessary part of your deployment scheme.

WATER DELIVERY

Once you have determined that you have a solid water supply, and have selected your attack mode, then comes the task of water delivery. Water can be applied in a number of ways:

- Attack hoselines: 1-¾-inch, 2-inch, 2½ inch

- Master stream devices: fixed, portable, or elevated

Please note that in my list of hose above, no mention is made of a booster line. This little red bit of evil has no place in the modern attack arsenal. I recommend that you save money the next time you buy a pumper by eliminating this leftover contraption from the era of chemical fire engines.

5.6 Know where your water supply sources are and how to get the most from them. *Courtesy of Harvey Eisner.*

5.7 Fires do not always conveniently locate themselves near established water supply sources. Know how to get water by alternative means.

An attack line with a variable flow nozzle can deliver flows ranging from approximately 50 to 350 gallons per minute (gpm). And you can carry both adjustable and straight stream nozzles to provide for any eventuality. I really think that only two sizes of hose are needed on a modern attack pumper. These units should carry a stock of large-diameter supply hose and an inventory of initial attack hoseline. Remember that a 2-inch attack line can flow any normal amount of water that you will need to attack a fire or provide exposure protection in difficult-to-reach positions.

Master Streams

If your water needs become greater, you will have to resort to the use of master streams. As you know, the term master stream is used to describe any fire stream that is too large or powerful to be directed without mechanized assistance. The various types of master streams in use today include:

- Elevated streams (Figure 5.8)
- Turret pipes
- Monitors
- Deluge sets

Regardless of the type, each of these has the same basic purpose. They are designed to put a lot of water on a fire. The different uses of each come from the way you as the Incident Commander want to attack the fire. You should consider the following when deciding which type of master stream to use:

- Monitors can be fixed or portable (Figure 5.9).
- You can place a monitor where you want it.

5.8 Make sure that monitors are operated safely. *Courtesy of Harvey Eisner.*

5.9 It will be your decision as Incident Commander where to place an elevated stream. *Courtesy of Harvey Eisner.*

- The turret pipe is permanently affixed to a pumper with a direct connection to the pump.

- Deluge sets require close monitoring to ensure that they do not "walk about."

- Elevated streams can be both portable and fixed.

- Elevated streams can be both manual and automatic.

- You should never operate a master stream where it can punish your personnel.

Master streams should be located at the point of best advantage . Considerations for this placement should include:

- Stream power

- Reach

- Safety of personnel

Always consider the concept of collapse zone when setting up your master streams. Once streams are in place, they cannot be quickly moved. Always consider a minimal collapse zone to be an outward distance from the structure that is 1.5 times the height of the involved building.

The deployment of engine company resources is like moving pawns about on a chessboard. When you decide what type of fire you have and where it is headed, you can deploy the number of master streams and hoselines your experience tells you are necessary. Here are some observations that may help you in deploying your fireground engine company resources:

Carter's Engine Company Tips

- **Most fires you will ever attend can be controlled by a single, well-placed hoseline.**

- **Do not shoot water at smoke.**

- **An effective hose stream will usually extinguish all the fire it can in less than a minute.**

- **Keep your hoselines moving until the fire is out.**

- **An effective interior fire sector commander will ensure that lines are being moved in on the seat of the fire.**

- **The Incident Commander and subordinate sector commanders must work to prevent hoselines from working against each other.**

- **Never stretch light and hope for help from the next company.**

- **Whenever possible, back up your initial attack line with a line of at least the same size.**

- **Always call for help early. If your tactics are not working, try something else.**

- **Avoid poor use of personnel (such as six people on a 1C|v-inch hoseline)**

- **Do not let a smoke condition scare you out of making an interior attack**

These are the basic rules I have used for the better part of the last 30 years. Some are so old that they have grey hair. Use them nonetheless. A lot of fine people spent their careers making sure that these suggestions worked. I offer them to you to take to future generations.

6

Truck Company Operations

At every fire, there are a wide range of functions that have nothing to do with the application of water to the seat of an uncontrolled fire. These so-called **Truck Company Operations** are performed by aerial ladder companies in the larger cities. In smaller communities, specially-equipped pumper companies or squad units carry the tools and possess the talents to get the work done. Truck company work is one of the most frequently overlooked of the tactical fireground functions, yet proficiency in all these functions is necessary if lives and property are to be saved.

Included in this discussion of truck company operations are some tips that I have learned through decades of hard service. They are offered as a starting point for your learning curve. You should apply them and revise them as your experience dictates. They are basic preventive measures that will help you to do your job safely. I call them **Carter's Truck Company Tips**.

LIFE SAFETY

Your number one priority at every alarm is the saving of lives endangered by fire. In order to do this, however, you must see that your firefighters operate in a safe and organized manner. Not only is it potentially disastrous if firefighters become injured or lost, it also requires that additional firefighters be deployed to help their fallen comrades.

Truck Company Tip #1 is very basic: **Human life is your primary concern**. And this concern starts with you and moves outward. Your first concern is with the safety of the fire fighting crew entrusted to your care. They must be nurtured and preserved. Then comes the endangered population of the burning structure you have been called to. Lastly, you must have a concern for the general populace of the area surrounding the fire scene. You should not kill innocent citizens just because you want to drive fast or operate carelessly.

It is difficult to provide hard and fast rules about when, and more important, when not to enter a burning building. In far too many cases, people are endangered in heroic but foolhardy rescue attempts. No one likes to play God. However, there are certain situations when there is no reason to waste your most important resource: your people. If your brain is telling you, "Nobody could live through that," pay attention to your instincts (Figure 6.1).

Personnel safety is the fuel that powers the decision-making engine in life and death scenarios. This is not to say that an aggressive interior fire attack should be avoided. Training, experience, and circumstances dictate when to make that "Cavalry Charge" up the front stairs of a burning building (6.2).

6.1 Do not commit troops when it obvious that there is no possibility of rescue. *Courtesy of Harvey Eisner.*

6.2 Training, experience, and good on-scene evaluation will give you the best clues for deciding whether an aggressive initial attack is prudent. *Courtesy of Ron Jeffers.*

When you are evaluating life safety issues, remember the following:

- Seek to save the lives of those most endangered first.

- Those who yell the loudest are not necessarily in the greatest danger. It may be the unconscious person in the midst of a cloud of boiling smoke who needs your help first.

- Forces may need to be diverted from other fire fighting operations to save lives.

- Generally speaking, you should allocate a two-member search and rescue team for every 2,000 square feet of property to be searched.

POSITIONING/RAISING LADDERS

Truck Company Tip #2 deals with apparatus positioning. **First-due truck company takes the front of the building, or best point of vantage**. A well-equipped aerial ladder company does no one any good when it is parked around the corner, or on the wrong side of the building. Your people should have easy access to the various pieces of specialized equipment available on a truck company. If conditions allow, you should raise an aerial to the roof of the fire building. Bear in mind that you should never challenge high-tension electrical lines for any air space in the vicinity of a burning building. There is no justification for slipping aerials through overhead electrical wires. This can have fatal consequences (Figure 6.3).

Truck companies are called upon to raise ladders to aid in search and rescue, perform ventilation, initiate fire attack, and to ensure that firefighters have good access to escape routes at all levels of the fire building. Some of the things you must consider are:

- Aerial to roof as it is fairly easy

- Ground ladders as needed for the fire floor, as well as at least one floor above and the floor below the fire.

- Small ladders inside building.

- Special ladders as needed on the outside.

- Keep a proper distance from electrical wires.

SEARCH AND RESCUE

Truck Company Tip #3 deals with search and rescue. **No one goes in alone**. The buddy system is the basis for all that we do on the fireground. At least one member of each team should have a portable radio. And all members should wear and use their PASS device. What good are you if you are moving in alone and are knocked unconscious by a falling ceiling?

There are guidelines that tell us the minimum crew size for fire fighting operations. Remember that the Federal Occupational Safety and Health Administration (OSHA) requires that a four-person team be assembled before beginning interior structural fire fighting operations. And rescue is a very labor-intensive operation. Many times it takes two rescuers to rescue one victim. Two people are needed to properly search a room so that all parts of an area are covered. For more on this topic, review the latest edition of **Essentials of Fire Fighting** or **Fire Service Rescue** from IFSTA.

6.3 Always keep in mind the location of electrical wires when positioning aerials. *Courtesy of Harvey Eisner.*

It is essential that search and rescue operations be performed in a thorough, careful, and methodical manner (Figure 6.4a and b). If you miss an area, people can die. Some very important safety guidelines for conducting search and rescue operations are as follows:

- Work in teams of at least two members.

- Each team should have a radio.

- Start at the door (outside and inside).

- Work inward toward the center of the fire building.

- Stay along the walls, occasionally crisscrossing the hall or room you are in to check the center area for victims.

- Check under tables, behind furniture, in bathrooms (tub and shower stalls), under beds, and in closets.

- Vent as you go, when practical.

Search Priorities

6.4a You must establish a set of priorities before you search.

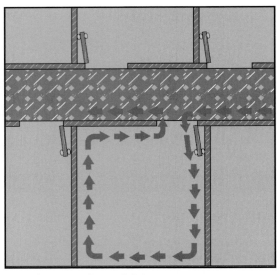

6.4b Always search in a definite pattern.

- Note escape routes, windows, and doorways to other areas.

- Continually communicate with your partner

- Stay within sight, touch, or voice of your partner.

- If you find a victim, consider the quickest way out, call help as needed, and ensure that the search is completed.

Truck Company Tip #4 also deals with search and rescue. **Once is never enough** when it comes to searching for potentially trapped fire victims. You must ensure that both primary and secondary searches are conducted. Many times only a quick look is taken into the fire area. And in other situations, an unorganized search is performed in a hurried fashion. There must be a method to our search and rescue madness.

You should be on the lookout for victims as soon as you enter the fire building. These people might have passed out just short of the door or window. You can remove these people quickly and return for more. Your portable radio can be used to get help or direct others to previously unsearched areas. Always consider the quickest way out for the victim. Remember nonbreathing victims have but a few moments before brain damage and death.

Truck Company Tip #5 might seem obvious, but is often overlooked in the haste of an emergency scene. **Search in a definite pattern**. It is critical to emphasize that your search must be conducted according to predetermined guidelines. All floors that are involved in fire must be examined.

The most essential element in a search and rescue operation is time. The more time that the fire allows you, the more people you can save. Here are some things to remember:

- Be sure someone has a line working on the fire.

- If you cannot see your feet, drop to your knees and proceed by moving forward on your hands and knees (Figure 6.5).

- By using a tool to feel ahead of you, "sound" floors, halls, and stairways before entering.

- Completely search one room before moving on to the next.

- Start your search on an outside wall. You can ventilate as you move about the room.

6.5 To learn to operate in smoky conditions, firefighters should practice searching on hands and knees.

- Be sure to search under all furniture; move it if you must.

- Mentally note the location of emergency escape routes.

- Search under beds, in bathtubs, and shower stalls.

- Periodically stop and listen. Sometimes you can hear people moaning or calling for help.

- Report any extension of fire

- If heat prevents you from entering a room, feel ahead through a door or window with your tool. Maybe you can locate someone who is within easy reach.

- Once you have removed a victim, place the individual in someone's care so that they will not go back into the building. This has happened and people have died.

Have a means for marking an area once it has been searched. This is particularly important in larger occupancies, both residential and commercial. Some fire departments carry pre-printed tags. Others use chalk or other temporary marking devices. One of the old marking standbys involves turning a chair over near the entrance to the room or area. Whatever methods you use, be sure that you have a plan for conducting your search and rescue operations.

During the course of a search and rescue operation a firefighter might become disoriented or begin to lose his or her cool. The following are some further tips for completing your mission safely:

- If you start to feel lost or lose contact with your partner, get to a wall and follow it to the door or window.

- Attempt contact with your partner.

- Follow the hoseline. You will either meet the attack team or find your way out.

- If you can, get to a window or safe place and call for help.

- Let your officer know your partner may be missing.

RECONNAISSANCE

A well-trained truck company can make the Incident Commander's job easier. They can identify problems inside the fire building that might be unknown to the command personnel. They are like the Combat Engineers in the military, who go in ahead of the main force and lead the way, removing obstacles for the commander and his force. In the same manner, truck company members must report back to the Incident Commander on each step they take to open up, search, and ventilate. Members of the unit should search for hot spots and fire extension as they move through the building. Anything that seems out of order should be communicated to the Incident Commander.

FORCIBLE ENTRY

A truck company is required to perform forcible entry as needed at the scene of an emergency. A number of different areas must be addressed. Before firefighters can enter a building, they may have to force entry. Whether it is through a door, window or other opening, it is the truck company's job to see that it is done (Figure 6.6). Entire fire fighting operations have been stymied because there were no people to force entry. Something as simple as opening the right door at the right time can save a life — or prevent the destruction of a building.

Remember that if you give your fellow firefighter a way in, it is always good practice to open a secondary way out. This is done in case you lose the original way in, for whatever reason. Let us look at the various entry points you may be called upon to breach:

Doors - There are four general types that must be considered. They are revolving, overhead, sliding, and swinging. Each has its special characteristics and a fire department must practice with each.

Windows - The same holds true for windows. There are a number of different types, each with their own particular nuances. Know them. And believe me when I say that practice in this area does make a difference.

- Check Rail
- Double Hung
- Louver

- Jalousie
- Factory

Remember the simple statement in the IFSTA's **Forcible Entry** manual, which tells us to always, "try before you pry." You do not necessarily have to break things. Be inventive and be non-invasive. However, if time is of the essence, "**in is in**," regardless of how you accomplish the task.

VENTILATION

Correctly performed ventilation can save lives, reduce the extension of fire, and make the attack easier. In order to see that the job of ventilation at a fire is performed correctly, you should adhere to the following guidelines:

- Be sure that fire attack personnel are ready to deliver water.

- Ventilate from the top down

- Ventilate from leeward to windward

- Crew of two to the roof position

- Officer and one/two firefighters to interior

- Use existing openings on the roof as needed: doors, scuttles, skylights, vents
  - create your own with the saw/use pike pole to push down ceiling.

- Open the windows on leeward side first

Truck Company Tip #6 can be a breath of fresh air for your fire fighting operation: **Vent high, vent low, vent often**. One of the major problems encountered during fire fighting operations involves smoke and heated fire gases. If not properly handled, they can be fatal. At the very least, they can slow an otherwise effective fire fighting attack.

Vent high involves creating openings in the roof. This is not the simple task it once was. Newer lightweight truss construction has created roofs that will not stand up to the weight of a firefighter equipped with a full array of personal protective equipment. The best way to know what types of construction exist in your area is to visit and pre-incident plan your response area on a continuing basis (Figure 6.7).

6.7 Lightweight construction can make roof operations especially dangerous. Know ahead of time how the buildings in your area are constructed.

6.6 It is the truck company's job to force entry if necessary. Make sure they know how to do it as safely as possible. *Courtesy of Harvey Eisner.*

The manner in which one would vent an old three-story frame tenement in an urban area might be quite different from a modern townhouse or condominium. The same comparison could be made between a new lightweight steel warehouse and an old fire-resistive concrete monster. Many of these older buildings are simply too tough to fall down.

Because of the problem with modern construction, a great many fire departments have turned to the concept of positive-pressure ventilation. Here is our **vent low** statement. Rather than creating openings in a structure to let the smoke out, high-pressure fans pump air into a fire building. This approach shows a great deal of promise. However, it must be used carefully or it will spread the fire. You must be ready to move in quickly and take advantage of the effects of the positive ventilation on the smoke and fire conditions. And you must have enough fans to do the job.

Vent often means to take any opportunity to allow smoke out when you can. As you move through the building on your search pattern, pause and open windows wherever possible. If you are in a position to use roof ventilation, create one large opening (4' x 6' min.). This will have a better effect than a bunch of small holes. And there will be fewer chances to fall through the roof if there are fewer holes.

And do not forget the old horizontal school of ventilation theory. Open windows on the leeward side first and then move to the windward side. This will allow a natural cross movement of air. You open them in the leeward/windward fashion so that there is not a sudden buildup of air. This may stimulate a backdraft if all other conditions are right.

SALVAGE, OVERHAUL, AND MORE

Truck Company Tip #7 is simple. **When you think that you are done, there is always one more thing to do.** We have already addressed rescue, ventilation, and laddering. Other tasks that you may need to consider are:

- Checking for fire extension (Figure 6.8)

- Salvage

- Overhaul

- Utility control

- Ladder pipe operations

- Property conservation as needed

Salvage has become the stepchild of the fire service. Many years ago we took pride in this task. How many of you can remember being taught to leave the building in better shape than when you arrived? This is not necessarily the case any longer. Nonetheless, you should strive to preserve as much property as possible.

6.8 Performing overhaul is necessary if you don't want the fire to come back at you again. *Courtesy of Harvey Eisner.*

The primary problem comes from the fact that water is our weapon of choice. But effective salvage operations can reduce the damage toll. You can cover belongings, remove water from the building, or remove endangered property to the exterior. Another aspect of salvage that is frequently overlooked involves removing water from the fire building to prevent structural damage. Once again, you should turn IFSTA's **Loss Control** manual for guidance on this topic.

The truck company is a critical component to leaving the building in a safe condition. They do this via the overhaul function. A truck company must overhaul in order to:

- Prevent rekindle
- Preserve any evidence of arson that might be found
- Do as little damage as possible
- Protect personal belongings

Ceilings, floors, and walls must be opened up as needed. This will uncover the hidden fire that could spread unchecked if it is not found by the truck company. A number of very good stops have been made by fire fighting crews that have opened up the fire building in order to cut off the fire. Many firefighters have been killed or injured because they were unaware of the presence of fire in a hidden area until it was too late. One of the primary objectives of a truck company is to "get it open."

Finding the fire can be a primary or secondary task for a truck company. You may well find some fire extension as an incidental sideline to search, rescue, ventilation, and forcible entry. Or it may be your primary task during overhaul operations. You should be aware of where the nearest hoseline is. You must be able to call for water whenever and wherever you need it. Prompt fire extinguishment can save a lot of time and property. A competent truck company can assist in this task.

Do not overlook building utilities during your fireground operations. A team of two firefighters should be ready to shut off gas, water, and electric. This team should be equipped with a portable radio to maintain contact with command.

Fire personnel should be also trained to look for those clues indicating that a fire may have been intentionally set. Such things as multiple points of origin, excessive structural damage, and familiar aromas, among other things, should be noted and passed up the chain of command so that it can reach the proper investigative group. This training and experience will be particularly valuable during the overhaul phase. For further information, consult the IFSTA's **Fire Cause Determination** and **Fire Investigation** manuals.

Property conservation is an operational area frequently overlooked by suppression personnel. Steps should be taken to ensure that unnecessary damage is not done by operational personnel. Covering furniture and personal property with tarps and salvage covers can prevent water damage. De-watering operations include portable pumps, water vacuums, water slides, and structural alterations that shunt water out of the building. People will remember that you took care not to do excessive damage to their property.

A full set of IFSTA manuals is critical if you are to gather all of the necessary knowledge you will need to be an effective firefighter and fire officer. Learning what you need to know how to do your job is truly a life and death affair in our line of work. A review of **Carter's Truck Company Tips** is as follows:

Tip #1: Human life is your primary concern.

Tip #2: First-due truck company takes the front of the building, or best point of vantage.

Tip #3: No one goes in alone.

Tip #4: Once is never enough (especially for search and rescue).

Tip #5: Search in a definite pattern.

Tip #6: Vent high, vent low, vent often.

Tip #7: When you think that you are done, there is always one more thing to do.

7

Pre-Incident Planning/Sizeup Considerations

Fire officers have long struggled to evaluate and prioritize all of the information they need to operate safely at the scene of a fire. Over the past six decades, numerous fire service researchers have labored to quantify the clues that a fire officer needs to observe to make good decisions on the fireground. The tactical deployment of fire resources, in response to the demands of a particular situation, is a basic component of fire fighting operations. A systematic review of fireground conditions from arrival to departure is a skill that all fire officers must master. Because there are a number of important factors to be considered, it is the intent of this text to group the factors into **sizable memory bites** in order for them to be better utilized on the fireground. My experience is that the more information a fire officer can call up from memory during the stress of an emergency, the better will be the quality of his or her decisions. This is especially true for strategic and tactical deployment decisions.

In this day of the Incident Management System, people have inadvertently forgotten what it takes to fight a fire. Some people would have you believe that once the sectors are in place and all the tasks delegated, that the pieces of the entire operational puzzle will magically fall into place. Unfortunately, this reasoning misstates fireground reality. Whether you use IMS or not, you must still ensure that the classic elements of Rescue, Exposure Protection, Confinement, Extinguishment and Overhaul are performed by your firefighters.

Critical to the performance of each of these is proper sizeup of the emergency scene, and critical to proper sizeup is appropriate pre-incident planning. The command officer arriving at the scene (of a fire) must quickly gather information needed for intelligent decisions. Sometimes there is not as much information as we would like. Regardless of the amount available, it must be analyzed. It must then form the basis of your action plan for fire attack. This task must be done in an orderly fashion. In this way, you will limit gaps in information and conditions. To ignore the use of a sizeup is to invite disaster.

Over the years I have studied a number of sizeup lists. Each textbook treats the topic just a bit differently. Based on my research and experience, I have come to use certain factors at every fire I encounter. A quick look at this list is a critical starting point for the new approach I will introduce in the next chapter. I always look at:

- Life hazards
- Time factors

- Structural height constraints and demands

- Area problems

- Building construction type

- Occupancy type and associated hazards

- Location and extent of fire upon arrival

- Exposure problems

- Auxiliary appliances available to assist fire department operations

- Available water supply

- Wind and weather conditions

- Apparatus needed to operate and available for use

- The level of communications equipment needed to confine, control, and extinguish the fire

- Salvage and overhaul

- A whole host of special matters that must be evaluated for their impact upon the basic tasks needed to combat the fire: things like hazardous materials, structural collapse, and special equipment situations.

This list is extensive. However, it does constitute the basis for those tasks that the Incident Commander must accomplish. It is my intention to take a look at these factors and then to introduce a new view of fireground sizeup.

LIFE HAZARDS

The first priority at any fire is the preservation of human life. The only time that firefighters should be placed at risk is when there is a probability that human life can be saved. In those cases where people in a burning building are obviously beyond help, do not jeopardize the firefighters (Figure 7.1).

The primary indicator to guide you in this decision is the volume of fire showing upon arrival. It is a fact that a given volume and intensity of smoke and fire kills people. Look at what you have to contend with and risk your people only when the opportunity for a successful rescue presents itself. This is an observation that requires education and experience. Do not add to any existing toll by needlessly sacrificing fire department personnel. Ask the following questions with regard to life risk at a fire:

- Are people trapped? (Figure 7.2)

- Where might they be trapped? (Fire floor/above/below)

- How many people could be trapped?

- Are there exits that can be secured to assist with the evacuation of endangered people?

- What are the hazards to firefighters?

7.1 The primary indicator of possible live victims is volume of fire showing on arrival. *Courtesy of Harvey Eisner.*

7.2 Top priority at any fire is rescuing people safely. *Courtesy of Harvey Eisner.*

While the answers to these questions are usually available, it may be necessary to search for the answers to questions regarding available exits, hazards to firefighters, and number of people trapped. Remember, the only way to be sure that every possible live person has been rescued is through the use of a thorough search and rescue operation. Experience has taught me that bystanders can generate faulty information. Nonetheless, you should still ask people in the vicinity of a fire if anyone is still in the fire building. Weigh this information carefully and weave it into your decision-making process.

One question that must be asked early in a fire fighting operation is whether or not a quick hoseline can cover your rescue route. I have encountered situations in which an escape route for occupants was threatened by fire. In many of those cases, a hose stream directed between the threatened egress and the fire made the difference between a safe escape and death or serious injury. This is advice that has been forgotten or ignored by a growing number of fire department personnel, in pursuit of the spectacular rescue. Many fine rescues have been made as a result of a hoseline holding the fire out of a staircase while the people exited.

TIME

Time can have a direct bearing on the number of lives that may be at risk from a fire. Always consider the following elements at a fire:

- Time of day
- Time of week
- Time of year

Each of these elements can tell a great deal about whether a structure is occupied or not. In many cases, the answers to time-related questions are so simple that they are overlooked. Is a school on a Monday afternoon in January likely to be occupied? Would this be the case on an early August evening? Do certain traffic bottlenecks exist at various times during the day? Allow for these considerations in response routes, as well as in any necessary calls for additional assistance.

How about weather? It takes extra time to arrive at a fire during the snowy months of winter that a lot of you face every year. It also takes more time to mount and pursue your attack (Figure 7.3). Operations must proceed more

7.3 Operations at the emergency scene will be significantly affected by severe weather. *Courtesy of Harvey Eisner.*

slowly and cautiously when ice is present. Are hurricanes a problem? They can place severe demands on your operation. Normal routes of response may be unavailable. The time necessary to stage and deploy resources on the fireground will be protracted. And resources normally included in your mutual aid network may be tied up with operations in their own communities. These are real-life problems.

What about holidays versus normal days? Would a fire in a shopping center on Saturday, December 24 give you more to worry about than one on Tuesday, May 27? You bet! And you must allow for all such considerations in your sizeup of the fire situation.

HEIGHT CONSTRAINTS AND DEMANDS

The height of the building can pose some serious questions for you (Figure 7.4). If a high-rise building is involved, do you know where the stairways are and where the safe areas are located? Where are the elevator banks? And where are the hidden spaces? If it is a normal, low-rise building, can you ladder the building? Are there obstructions? (Figure 7.5) Do you have enough firefighters to perform any necessary ladder work? These and other questions regarding height must be answered in order to mount a successful fire fighting operation. This is an area where information gathered during pre-incident planning sessions can come in very handy.

AREA PROBLEMS

Another factor that can be determined ahead of time during planning sessions is the matter of **area** of the building in question. Is it large and open or is it narrow and compartmentalized? Each has certain things to worry about. In a large open area, everything is usually exposed to the heat and destructive force of the fire. There is a greater heat buildup. This then can lead to a larger fire, more structural damage, and more hazards to your personnel. Greater heat and punishment may have to be endured to gain your objectives.

7.5 Even an ordinary structure can have obstacles, such as these trees, that make fire fighting operations more difficult.

7.4 Many more personnel will be required to combat a high-rise fire. *Courtesy of Harvey Eisner.*

Certain types of buildings almost always succumb to the ravages of fire. Churches, supermarkets, and bowling alleys have classically been the scenes of more total loss fires than others. You are usually dealing with large open areas, large hidden areas, and parts of the building that are inaccessible to easy fire department entry. Do not kill or maim people by placing them in needless danger at these types of buildings. Remember the truss roof that collapsed at the supermarket fire in Brooklyn, New York back in 1978, or at the Hackensack tragedy a number of years ago. These tragic lessons came at great personal cost. Learn from them.

In those buildings that are smaller in area, you face another group of problems. You face greater search and rescue requirements because of the large number of areas, each with its own individual place to be checked. There can be many hidden areas. Fire is usually a lot harder to locate under these circumstances. However, it is usually easier to move in and knock down a fire in a smaller area.

CONSTRUCTION

Construction is a factor that comes into play at every fire. It breaks down into two basic areas of consideration: those where the building materials can help you, and those where the building materials make things worse. Buildings constructed from materials that are somewhat fireproof in nature are an ally of the firefighter. Brick and masonry afford a degree of protection; however, this is usually negated by the newer forms of structural members. Construction that features trusses stapled together with steel gussets and wooden I-beams, held together by glue, gives me nightmares. And they are now favored by the people who make their livelihood in the construction business (Figure 7.6).

Fires in older, balloon-frame structures can spread rapidly. If a fire seems to be moving up a wall faster than you think it should, you are probably dealing with a balloon frame. Work to get ahead of the fire by anticipating where it will spread.

Steel and iron I-beams conduct heat very quickly. Protected steel provides a measure of safety because the steel is coated with a fireproofing material. A serious fire will eventually cause this to fail, but you have a measure of protection over plain steel. Iron beams can carry heat, fail very quickly, and pose a great deal of danger to operational personnel.

Remember that all forms of masonry eventually fail. As the heat of the fire drains the moisture away, masonry members approach the point at which they will decompose. This may be explosive, as in the spalling of concrete, or it may be slow, as in the crumbling of sandstone. Either way, it poses a danger to you and your firefighters.

Pre-incident planning inspection visits are the perfect time to gather the construction information you will need for the sizeup process (Figure 7.7). Remember the old maxim, **"know it before you need it."**

OCCUPANCY TYPE

A study of the occupancy type in any fire scenario can give you a lot of clues as to the hidden (or not so hidden) hazards you may encounter at a particular building fire. Residential fires present life hazards as the primary concern. However, what about all the household chemicals you might find in a kitchen or the flammables in a handyman's workshop? I can remember a severe injury that came from exposure to a fire involving pool chemicals; the man involved almost died. Anything can happen at a house fire today. Be aware and be wary.

7.6 When you know that certain types of construction, such as these lightweight roofs, will make your job harder, adjust your operations accordingly.

7.7 There is no substitute for the pre-incident planning visit to learn about the hazards that may be inside a structure. Note the paint booth inside this facility.

Every industry has its share of user-specific hazards. One of the toughest fires I ever encountered involved the dip tanks used in a metal coating company. The material was extremely flammable and toxic and the sprinklers in the building caused the material to almost overflow its containers.

The name of a company can give you an excellent initial clue as to what you are facing. How about the XYZ Chemical Company? What might be the problem at Al's Paint Supply? The ABC Feed and Fertilizer Store? These and similar clues might be your only immediate chance to keep your people from being injured. Pay attention and pre-plan.

APPLIANCES AVAILABLE TO ASSIST

A number of installed fire protection appliances exist to help the poor harried firefighter perform the task of putting out fires in large buildings. Each of these devices has a place in strengthening your fire fighting attack. Use them to the best possible advantage and make the task easier for your firefighters (Figure 7.8). Their location and availability can best be determined during your pre-incident planning visits:

7.8 Know which features of a building, such as self-closing doors, will assist in compartmentalizing a fire.

- Solid masonry fire walls

- Self-closing fire doors

- Roof parapets

- Fire-resistive stairway enclosures

- Fire alarms: to alert you and get the people out

- Automatic sprinklers: for rapid interdiction

- Standpipes: Your fire mains in the sky at high-rise fires (also good for large-area structures) (Figure 7.9)

AVAILABLE WATER SUPPLY

Nothing can happen at a fire until the wet stuff hits the red stuff, so water supply is a crucial element in your operation. You must know which water mains can deliver the greatest fire flow. In this way you can deliver the optimum effect from your pumpers. Be sure you can provide the answers to the following questions:

- Are you on a grid system or dead-end main?

- Where are the hydrants?

- Are they high pressure or low pressure?

7.9 Your operations may depend on use of standpipes, so know where they are.

- Are there any special pumping features that can boost your pressure during emergency operations?

- Is water available in adjoining properties?

- Are sufficient tankers available should you need them?

- Are there private systems for your use?

- Might you have frozen fire hydrants?

There are some additional considerations for those of us in areas not blessed by hydrant protection. Think about the following:

- Where are my drafting points?

- How quickly can water tankers arrive?

- Will it be possible to develop a tanker shuttle?

- Can we relay water from a nearby static source?

- Will portable water tanks be available? (Figure 7.10)

7.10 You will not be able to successfully establish a relay shuttle if you have not practiced it.

- How long will it take to establish a viable water supply?

At 3:00 a.m. on a cold winter's night is not the time to look for water supply sources. This should be an on-going consideration in your inspection and planning programs.

WIND AND WEATHER

Wind and weather are two variables that you cannot, in general, plan for. It might be hot when it should be cold or vice versa. However, if you develop a general operating profile for any extraordinary weather that you might expect to encounter, it will be easier to work around when the time comes. How about some cold weather standard operating procedures (SOP's)? Such things as additional equipment and manpower might be crucial in muscling hose over snow-clogged streets. The same would hold true for hurricane, tornado, and monsoon conditions. You should determine the various weather-related hazards you might face by again asking yourself, "*What if...*? You must then develop procedures for operating under these situations. I have fought fires in chest-deep water as well as waist-deep snow. Neither was much fun, but if you are prepared by experience, education, and training, the job can be done. You should take the time to consider what extremes of weather may require. And then prepare for those eventualities (Figure 7.11).

Once the fire occurs, take time to note the wind force and direction. This can play a great part in the potential for exposure protection. Dry weather places added burdens on you; so does wet weather. Be aware of your surrounding climate and be able to react accordingly.

COMMUNICATIONS

It has been my experience that everyone wants to talk at once during a working fire. A system for channeling communications must be developed as one of your standard operating procedures (Figure 7.12). You should also ensure that sufficient fixed, mobile, and portable communications capability exists. It sure is difficult to rustle up a spare radio at 3:00 a.m. unless a supply is available from central point in your department. You may choose to expand your capabilities by becoming part of a regional communications system. Only you know what is needed in **YOUR** community.

SALVAGE/LOSS CONTROL

Salvage is an area of fire department operations that most often receives little more than lip service from the powers that be. In order to provide the fullest possible property conservation service to your community, you must provide sufficient salvage capability. It should arrive in a timely fashion and be provided by personnel trained to know the difference between a tarp and a salvage cover. If the emphasis is not provided during the training, equipping and planning phases, proficiency will not appear on the fireground (Figure 7.13). For an in-depth consideration of this often-neglected aspect of

7.13 Salvage is an often-neglected aspect of fire fighting, but one that can be very valuable to a building's occupants. *Courtesy of Ron Jeffers.*

7.11 If you think you will need more help, call for it immediately. *Courtesy of Ron Jeffers.*

7.12 Your communications systems must be functional and well-rehearsed if they are to be effective. *Courtesy of Ron Jeffers.*

fire fighting, consult IFSTA's **Loss Control** manual. Again, pre-incident planning and GOG's will give you critical information about steps that you may be able to take to minimize loss, such as covering expensive equipment or simply moving certain items outside immediately.

ANCILLARY FUNCTIONS

There are a number of ancillary functions that form a part of your fire fighting operation. Although they are not strictly fire department functions, you ignore them at your own peril. These organizations should be a part of your standard pre-incident planning operation:

- Police will be needed for crowd and traffic control. They may also have a role in the investigation phase of your operation, depending on local policy.

- You will need support from the electric, gas, and water companies. In some places they are all municipally owned. In other places they are privately owned. You should work to interact with these folks on a frequent basis. It will pay dividends on the fireground.

- Local Office of Emergency Management officials can provide a wide range of assistance from outside sources.

- The road department can assist with heavy equipment, road barricades, and other support functions. Work out your needs ahead of time and make them a part of your pre-incident planning process.

- Salvation Army and Red Cross can provide disaster relief, as well as logistical support for emergency personnel operating at large incidents.

ON THE FIREGROUND

The answers to the next series of factors can only be determined at the time of any incident. When you arrive, what is the location of the fire and how extensive is it? You must develop answers to all of the following questions:

- Where is the fire?

- How bad is it?

- Is it near any vital stairway?

- Can your forces hold the stairs with an aggressively deployed attack hoseline?

- Can you secure a quick rescue path with a hoseline?

- If it is a cellar fire, is it a threat to the whole building?

- Does the time of day when you respond affect the occupancy on the upper floors of a residential building? High-rise? Commercial or mercantile building?

In addition to these thoughts, you must remember that when dealing with a cockloft fire, **"you've got to be in it to win it."** This is to say that the **carefully considered** use of an aggressive interior attack, where warranted, can provide marvelous results.

Location and Extent

Location and extent combine in a crucial equation. You can encounter a large fire in an open field and face very few problems. However, a very small fire at the base of the only stairway out of a residential home can be deadly (Figure 7.14). You must make this assessment every time you go to a fire. IGNORE THIS ASSESSMENT STEP AT YOUR OWN PERIL. By focusing on location and extent you zero in on the steps needed to control and extinguish a fire.

7.14 The location of a fire can be just as important as its size. This fire, at the base of a set of stairs, is blocking the only way out of a building.

Exposure Protection

As you are considering where the fire is, be ever mindful of where it can travel. This moves us into the realm of exposure protection. There are three sorts of exposures that you must be concerned with:

- Life
- Internal
- External

There are three distinct groups at risk in any discussion of life risk at a fire. The primary group for you to consider is your very own firefighters. If they are not available to perform their lifesaving work, others may die. They are also the life exposure most directly under your control. Next, you must consider the occupant population of any fire building to which you respond. Lastly, the general populace who may come to the vicinity of the fire must be controlled so that they do not become part of your problem. Calling police help for crowd control is an important consideration.

Another exposure problem you must address is the internal exposure. This might be any condition, such as duct work, conduits, open staircases, improperly installed curtain walls or the like which allows a fire to move within a building. Research and experience allow me to give you the following sage advice: "**The key to successful fire fighting is anticipation.**" The competent officer anticipates what might happen and takes action. Your favorite question should become, "*What if....?*"

The final exposure problem is the classic adjoining building scenario that comes to minds when we think "exposure." Just as anticipation of internal fire travel is important, so too is the consideration of travel from building to building. Factors such as size and intensity of the fire, wind and weather, and construction material of the exposed structure are all points to ponder. Please do so. Remember, there are six sides to any fire you are called to fight: **top, bottom, left, right, front, and rear.** Do not lose sight of any particular side lest the extension of the fire remind you of its proximity.

Apparatus Available

How much equipment do you have rolling in with you? Do you think it is equal to the task that you find in front of your eyes when you arrive at that fire (which always seems to come when you least expect it)? Can it handle what MIGHT happen, should your first cut at an aggressive interior attack prove unsuccessful? Do not look at where the fire is; rather, consider where it is headed. Then move to cut it off. You must consider location of the fire, extension probability, type and size of fire. Then drop back to a basic rule of fire fighting: **Big Fire = Big Water; Little Fire = Little Water.** Always act accordingly.

Try to keep a little something in reserve. It is a lot quicker to deploy a pumper at the scene than to wait for one to arrive, perhaps from a distance. Any pre-determined shortcomings in your department's response should be handled by automatic mutual aid. Get help rolling quickly. You can always send it back by radio if you find it is not needed. Better to be safe than sorry.

Special Considerations

Lastly, there are special matters that come up only infrequently in the course of a fire officer's career. It is their infrequent nature that can throw the curve ball at the unsuspecting fire chief. No matter how infrequently they occur, you must be prepared to respond to events as they transpire on the fireground. Some special matters for you to consider are:

- Explosions

- Structural Collapse

- Panic among building occupants

- A need for specially trained personnel or equipment
  - air crash
  - rescue
  - hazardous materials
  - cave exploration
  - farm equipment

- A need for large amounts of breathing apparatus

- Specialized extinguishing agents

- Poor terrain and approach

- Exits unavailable for anyone's use

SUMMARY

In this chapter, I have covered a number of important points that are considered by many in our field as crucial to pre-incident planning and initial sizeup. While it seems like a great deal to ponder in an emergency, all must be examined and worked through if you are to launch an effective operation. They have certainly guided me through many trying times. I am eternally grateful to the chiefs who came before me who had the foresight to commit them to paper. In this text, it is my task to take them into the future in a manner more easily usable on the modern fireground (see the next chapter). Take time to memorize your sizeup points. Learn the right way: it can save your life.

8

An Eight-Step Look at Sizeup

For more than 30 years now, attending fires has consumed large quantities of my life. It has been my pleasure to have done it on a number of different continents, for I have truly loved my work. And I have served a number of masters. I have been a military firefighter, a volunteer firefighter, a small town career firefighter and a big city firefighter. With this in mind, let me state the following as my basic philosophy of fire fighting:

A fire does not know what country it is in. A fire couldn't care less whether you are a paid, volunteer, military, wild-fire or industrial firefighter. Fire obeys some fairly universal physical rules. And fire travels in fairly consistent ways. Your job is simple. Learn all of the rules and apply them as necessary.

Having said these things, let me offer you some help in learning about fire fighting. You must assemble like fire fighting subjects and memorize them in usable groups. That is what I am going to do in this chapter. I am going to give you what will become known as **Harry Carter's Simple Rules of Sizeup and Command**. I have designed the format as a series of simple, easy-to-remember questions. These are the questions that I feel you will need to answer at every fire or emergency you will ever attend:

1. What have I got?
2. Where is it?
3. Where is it going?
4. What have I got to stop it?

5. What do I do?
6. Where can I get more help?
7. How am I doing?
8. Can I terminate the incident?

If you can answer each of these questions within the confines of the situation at hand, you will probably do well in the fire fighting business. Of course the Fire Gods will still toss you the occasional curve ball. But that is usually just to see if you are awake and paying attention.

These eight questions are designed to equip you with a standard decision-making process. While the questions seem simple, they are not. There is a depth and range to each one. However, if you can memorize these questions, you will be well on the road to success. For you see, the simple answers you will get on first cut will become better with age. As you fill in the first set of simple blanks, your mind will search for more facts and better information. In this way, you will be able to think your way through a fireground sizeup. Remember, rare will be the occasion when you can set up a lighted podium for your notes. You will have to use your brain, your instincts, and your Incident Management System control sheets. And believe me, the command sheets are going to be passive participants in the fire fighting operation.

This chapter will cover the basics of sizeup. The chapters that follow will build upon the basics, covering each aspect in depth. And then we will apply it to a variety of incidents that you may encounter. I want to get you thinking, to using these questions all the time. If you can answer these questions at a dumpster fire, then you should be able to answer them at a house fire. This is a standardized decision-making procedure that you can use time and again. And like all normal people, you will get better with repetition. Under each of these headings will come a series of questions that are specific to each circumstance. It is not my intention to establish a lot of hard and fast rules. In a fast-moving emergency scenario you need to be able to think and react quickly. That is the purpose of our question-and-answer mode.

I must pause and reemphasize a crucial point here: **answering these questions will be much more difficult if you have failed to perform the appropriate pre-incident planning**. In fact, you may find it impossible to answer some of them accurately if you are not well-prepared. Fighting fire effectively and safely will require all the tools in your arsenal; don't leave any of them at home.

QUESTION #1: "WHAT HAVE I GOT?"
Before you can go any further in sizing up a fire, you must firmly establish what you are facing (Figures 8.1 - 8.3). You do this because all other decisions come from this one. You must quickly decide what you are facing in terms of the following:

- **W** eather
- **I** ncident Type
- **T** ime Factors
- **H** eight Factors

Weather
Weather is basic to all that we do. There are problems with heat, cold, rain, and drought. Each has an impact on the manner in which you manage an incident. Unless you are experiencing severe weather, however, this area is a quick study: once you have decided that it is raining and the wind is blowing, you can move on to the next important area. Extreme or violent weather, however, will affect all that you do during the emergency response. It takes extra time to arrive at a fire during the snowy months of winter that a lot of you face every year. It also takes more time to mount and pursue your attack. Operations must proceed more slowly and cautiously when ice is present. Are hurricanes a problem? They can place some severe demands on your operation. Normal routes of response may be unavailable. The time necessary to stage and deploy resources on the fireground will be protracted. And resources normally included in your mutual aid network may be tied up with operations in their own communities. For those of you who respond in severe heat, remember that firefighters will have to be rotated out sooner and rehabbed to guard against heat exhaustion or heat stroke. These are real-life problems.

Incident Type

What type of incident are you dealing with? A number of other questions can then come quickly to mind:

1. Is there a life hazard? (ALWAYS OUR FIRST PRIORITY) (Figure 8.4)

2. Is it a vehicle fire?

3. Is it a structure fire?

8.1 "What Have I Got?" A house fire, so think about the necessity/possibility for rescue. *Courtesy of Harvey Eisner.*

8.2 If you answer is that you have a car fire, think first about your firefighters' safety. *Courtesy of Ron Jeffers.*

8.3 If you are facing a large fire, you know that you will need more resources. *Courtesy of Harvey Eisner.*

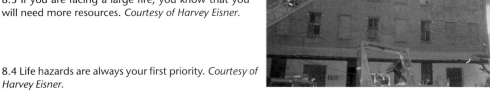

8.4 Life hazards are always your first priority. *Courtesy of Harvey Eisner.*

4. Is it a woodland or wildfire incident?

5. Is there something dangerous in the structure?

6. Is something really screwy involved?
   a. Oil storage yard
   b. Hazardous materials
   c. Chemical facility
   d. Piers and ships

Time Factors

There are a number of factors that vary by time of day, time of week, and time of year. Some of them are:

- School occupancy may be different at night or during the summer.

- Travel conditions in certain areas vary by time of day, week, or year.

- People usually sleep at night and are away at work during the day. However, in some areas there are many shift workers who are at home during the day. This is where knowing the demographic mix of your response area can come in handy.

- The number of workers in a given area may vary by shift time. This can be factory, communications, telecommunications, or any number of industries.

- A short trip in the summer can become a long trip during a winter snow storm.

- The holiday season can create heavy traffic jams in shopping areas.

You will have to survey your response area on a periodic basis to see how time factors can vary. Constant awareness of time-related issues is critical.

Height Factors

It has been my experience that fire suppression problems increase with the height of the building involved. Some of the things you need to consider include:

- Can you reach the roof of buildings in your district with ground ladders?

- Are all of the windows where rescue may be necessary accessible to your ground ladder complement?

- Are there areas in your community where you cannot operate your aerial devices because of overhead wires and/or narrow streets?

- Is your department trained and equipped to properly attack a fire located within a high-rise occupancy?

- Can your aerial devices reach all areas of an occupancy with expansive, modern landscaping?

- If you need aerial service and do not own such apparatus, where are they located and how quickly can they respond? (Automatic mutual aid can save precious minutes during an emergency.)

- Are your personnel trained to operate aerial and ground ladders? (Many fire personnel avoid ladders like the plague.) This can be a real problem, and it is one that can only be addressed ahead of time through periodic drills and assessment.

QUESTION #2: "WHERE IS IT?"

Once you have decided what you are facing you can then ask, "Where is it?" The fact that you have established what is the basis for where. In each of the situations listed above, you will need to determine where the incident is located. Where something is located is basic to reacting, attacking, and controlling the situation. Your basic concerns should be in the following areas:

1. Where in the world is the incident located?

2. In what part of your community or response area is it located? (Figure 8.5a and b)

3. Where in the vehicle or container is the problem occurring?

4. Where in the structure, or complex of structures, is the incident located? (Figure 8.6)

8.5a A fire in strip mall should immediately bring to your mind the need for certain resources.

8.5a A fire in an outlying area calls for a different type of response.

8.6 The location of a fire can be just as important as its size when it comes to planning your attack. *Courtesy of Harvey Eisner.*

Before you can decide on an appropriate course of action, you must be sure of the fire's location. How can you decide where it might go if you do not have a clue as to exactly where it is?

QUESTION #3: WHERE IS IT GOING?"

After determining the fire's location, you can then move on to the next question: "Where is it going?" This breaks down as follows:

1. Where is it going in the building?

2. Is it going to the next building?

3. Might it travel from a vehicle to a building?

4. Could it be moving from a contained position to an uncontained position?

It has often been stated that anticipation is the key to fire fighting success, and I subscribe to that statement. During my career I have seen numerous instances where the fireground commander lost the battle because he was three steps behind the fire. I want you to outthink the fire, so learn to ask yourself where the fire is headed.

If you do not take steps to protect exposures, fires will spread. This is a simple fact. A tremendous part of the question of where a fire is going involves where it is located and how big it is. **Location** and **Extent** of the fire on your arrival can tell you a great deal. The classic comparison could be stated as follows: Which fire is worse:

- A large fire in a junkyard (or)

- A small fire at the base of the rear exit from a two-story, two-family home?

My estimation is that the small fire is a greater danger because of its potential for the injury or death of occupants. The large fire would require a great deal of labor, but it should not be a life safety problem as long as you operate in a safe manner. However, a large fire in the center building in a row of similar attached structures can become a serious problem. By concentrating on exposures you can begin to limit spread. Once outward spread has been halted, you can concentrate on confinement and extinguishment of the fire.

It is important for you to remember that fire responds to the basic principles of any oxidation-reduction reaction. If there is heat, fuel, and oxygen available in the path of the fire, it will spread. Remembering the chapter of chemistry of fire, we know that heat can spread by conduction, convection, and radiant heat. Some simple facts relating to this are:

- Fire generally burns upward.

- Heat rises.

- Heat will rise until it meets an obstruction and then roll across it until it reaches a wall and moves downward.

- Heat can pass through metal ducts and pipes to spread fire.

- Old balloon-frame walls will allow a fire to spread upward quickly.

- Fire can squeeze through some real small passages.

Your knowledge of building construction, combined with your experience at fighting fires, will equip you with a feel for where a fire might be going.

QUESTION #4: "WHAT HAVE I GOT TO STOP IT?"

Before you can do something, you must be sure that you have sufficient resources to get the job done. This is why your next question should be: "What have I got to stop it?" This requires a fairly quick, but thorough analysis of your anticipated deployment force. You must assess the following areas:

- People and talents
- Apparatus and equipment
- Time to arrival of help

Again, here is where pre-incident planning comes into play: you already know the apparatus you have available and a standard response complement (Figure 8.7). You should already know available water supply and communications equipment. If the incident is at a structure, say a business, that has

8.7 You should know what types of apparatus you can call on. *Courtesy of Harvey Eisner.*

installed fire protection systems, you know what you will have to assist (or not). Do you have access to standpipes, sprinkler system, and so on?

By quickly asking and answering these questions, you can consider your range of attack options, as well as the potential for success for each alternative. After you have a handle on the possible movements that Mr. Fire can take, and you have your resources poised and ready, you are all set to "head them off at the pass." This quaint old phrase gains new meaning here. As a famous Southern general once said, "You have to get there fustest with the mostest."

QUESTION #5: "WHAT DO I DO?"

Having decided where you must go and knowing what resources you have to do the job, now ask yourself, "What do I do?" The range of alternatives is, fortunately, short. Your mode of attack can be:

- Offensive
- Offensive-Defensive
- Defensive-Offensive
- Defensive

Whichever is your choice, you must make it quickly and implement it as rapidly as safety concerns dictate. Many failures can be directly tied to the concept of "thinking a decision to death." Some Incident Commanders have more fun thinking than acting. This allows confusion to set in. And it is a perfect time for Mr. Fire to take advantage of the IC's inability to make a decision.

QUESTION #6: "WHERE CAN I GET MORE HELP?"

Whichever choice you make, do it quickly. At this point, you may need to rethink your position on additional help. You do this by consciously asking yourself, Where can I get more help?" These are the options for you to consider:

1. In your department

2. In the community at large

3. In the area

4. In the world

If you are a member of a large municipal fire department, it is incumbent upon you to analyze the need for additional alarms. Your needs can be slotted into the department's response scheme. As the size of an incident escalates, it may be necessary to request outside mutual aid (Figure 8.8). Remember that none of us is so big that we can do it all every time.

If you are a member of a smaller municipal fire department, your thoughts should quickly turn to your mutual aid network. Unless the groundwork has been laid during interdepartmental planning sessions, you may encounter problems meeting your needs. You would do well to work with your neighbors to set up a pre-determined Mutual Aid Box Alarm System. In this way you and the towns around you can structure pre-designated apparatus responses by street location. This system lets smaller communities request additional alarm support in the same way that major cities do. With this system comes certain economies of speed and size.

8.8 If you need a lot of help, call for a lot of help. *Courtesy of Harvey Eisner.*

Larger fires may require calls for assistance from further away. In many cases, apparatus and personnel must respond great distances to be of help to their neighbors. You must know ahead of time where the specialized help is and how long it will take to arrive. Anything that you can do or decide ahead of time will be of tremendous assistance during times of emergency.

It may be that for certain of the "screwy circumstances" mentioned earlier, help will have to come from faraway places. Hazardous materials scenarios can call for special skills that only a few firms or agencies can provide. Imagine the patience needed to handle a situation in Ohio, when the assistance team must respond from California. These things can and do happen, so you had best prepare for them. In order to handle unusual situations, your pre-incident planning must involve some fairly far-ranging considerations. This is critical for those times when you are forced to begin running through your checklist of questions. The more lead time you can build into any deployment decision, the better your chances of success.

QUESTION #7: "HOW AM I DOING?"

At various points during your emergency operation, you need to pause and ask yourself the next question: "How am I doing?" While you may not always want to confront reality, it is essential that you do. What are some of the ways that you can assess your success rate? How about:

1. If it was burning, has it been extinguished?

2. If the fire was moving, have you halted its spread?

3. If something was leaking, has the leak been stopped?

4. If a situation was dangerous, has it been made safe?

5. If something was bad, is it now good?

While these may seem like simple questions, each makes a point. And each forces you to think. Have you achieved the goals that you set for yourself during the operation? Has the situation stabilized? Are things really going your way?

It is not hard to imagine the relief you will feel when things go your way. Imagine that the major blaze that was burning merrily away in the center of a block of ten row houses is held to two in the middle. You had a fire, and the blaze was moving laterally. You decided to deploy your forces ahead of it. You then called for additional help and began an aggressive interior attack on the involved structures. Your goal was to limit the fire to the buildings involved. As you pause to drink a hot cup of java, you note that the fire has been stopped. So your answer to the question, "How am I doing?" will be a simple, heartfelt, OK.

QUESTION #8: "CAN I TERMINATE THE INCIDENT?"

At this point, you can ask the final question, "Can I terminate the incident?" In the case above you can answer yes. If what you wished to accomplish has been done, you can terminate (Figure 8.9). Each response will be a separate case. However, as your knowledge and experience levels increase, your ability to ask and answer these questions will improve.

8.9 If all goes well, you can terminate the incident successfully.

SUMMARY

This chapter has given you an overview of the fireground operating method that I have developed during thirty-plus years in the fire and emergency service world. Think of it as a combination of sizeup and strategy and tactics, all rolled into one. By asking and answering these eight questions, you will learn to think your way through every fire you ever encounter. Using a system to structure your thoughts will improve your chances for safety and success. You must use your knowledge continually or you risk losing it. And then what will you do on that cold, dark day in January, when you really need to know what you are doing?

9

The Basic Question: What Have I Got?

In this chapter we will take a closer look at the first of our eight Strategy and Tactics questions: "What Have I Got?" Although there are but four words to deal with, there are many elements to those four short words. Let us contemplate some of the answers that you must consider as you stand in front of that next raging inferno. The factors you must consider fall under four overall areas:

Weather

Incident type   (WITH)

Time factors

Height factors

WEATHER

The weather is an obvious element. Unfortunately, it is so obvious that it is frequently overlooked as an important factor in fireground operations. It is also important to note that weather can limit your tactical options, which may require you to modify your overall strategy. For example, high winds or lightning can limit the tactical deployment of aerial apparatus as well as ground ladders. Heavy snow or flooding can limit access to narrow streets and fire lanes.

Weather can be predicted to some extent. And generally speaking, certain types of weather are prevalent at certain times of the year. In order to be ready for a fire fighting operation, you need to consider weather-related matters during your pre-incident planning operations. People in cold-weather climates know how bad things can be (Figure 9.1). They know how to prepare their equipment and their operational techniques to operate in sub-zero temperatures or in the midst of a snowstorm. The same holds true for fire personnel who operate in extremes of hot weather (Figure 9.2). They know what to expect from dry weather, with high winds. Many have spent decades perfecting operations under these conditions.

Many of us live in areas where the climate varies all across the board. Summer conditions can be hot and humid or hot and dry. Winters can be mild and relatively free of snow. Or they can be brutally cold, with deep snow. It is important for you as an Incident Commander to factor these variations into your pre-incident planning system.

On any given day, be aware of the weather conditions. Note the wind force and direction. The simple act of spotting a flag waving in the breeze can tell you volumes about what an uncontrolled fire

9.2 Operations during hot weather must include attention to rehab. *Courtesy of Mike Wieder.*

9.1 The entire emergency scene is even more hazardous during cold weather. *Courtesy of Harvey Eisner.*

might do. If it is snowing, be aware of the additional response time caused by ice-covered and snow-clogged streets. Call for help early and be prepared to wait. It may be necessary to activate special weather-related mutual and automatic aid calls. Again, these can be factored into your operation during the pre-incident phase.

If it is raining, be alert to the possibility of lightning and take note of areas that might be subject to flooding. In some areas, flash floods are common. Being an effective Incident Commander requires that you be alert to the world around you. This awareness can help fill in the gaps during the crisis-filled early moments of an emergency.

INCIDENT TYPE

The next question you must answer deals with the meat of the matter: What type of incident are you facing? Your problems might be among any of the following:

1. Is it an outside fire?

2. Is it a vehicle fire?

3. Is it a contents fire?

4. Is it a structure fire?

5. Is it a high-rise fire?

6. Is it a hazardous materials fire?

7. Is it a woodland fire or an urban/wildland interface incident?

8. Is it something really unusual, that you may only encounter rarely?

a. Subway	e. Shopping mall
b. Elevated rail system	f. Lumberyard
c. Temporary structures	g. Aircraft
d. Piers, docks, and ships	

Approach these special incidents with extra caution. Safety must always be your operative rule. The less frequently you encounter a given event, the less likely you are to be fully trained in dealing with it. Attempt to relate the known of your existing knowledge to the unknown of the situation. If you feel uncomfortable about risking your people, evacuate and wait for the experts. By the way, maintain a long list of telephone numbers for those uncommon incidents listed above.

Outside Fire

Outside fires can fall into a number of categories:

- Trash can
- Dumpster
- Construction debris
- Trash in the street

The Incident Commander should be careful not to be too complacent with regard to these every-day operations. I can remember the dumpster fire that became a hazardous materials incident because someone dumped chlorine bleach and ammonia into the container prior to the fire. Luckily we were wearing our self-contained breathing apparatus and no one was injured.

Car Fire

At first glance, a car fire might not seem like much of a problem. The average firefighter will see scores of car fires during his or her career. However, it is just this frequency that can lead to complacency (Figure 9.3). And complacency is a dangerous commodity for fire people. What must you worry about at a car fire? In this order, your concerns are:

1. Your firefighters

2. Trapped occupants

3. The general public

Notice that I pay little attention to the car itself. The modern car will destroy itself in fairly short order. Even my nice

9.3 Even though a car fire seems simple to deal with, do not let complacency set in. *Courtesy of Ron Jeffers.*

old Chevy is a mass of plastic within a metal box. So your major concern is extinguishing the fire with a minimum of risk for your people.

Let me ask you another question. What is the first thing to fail in the average car fire? Of course — it is the plastic cable that operates the hood latch. So you will have to do a great deal of damage just to get under the hood, where most car fires begin. In addition, modern security devices have made most cars a veritable fortress to enter. Of course, we will do our best to overcome these devices, which means doing a great deal of damage to the car. Even under the best of circumstances, we leave a car in far worse shape than we find it. So let us concentrate on approaching these incidents in as safe a manner as possible. If the answer to your "What Do I Have?" question is "a car fire," you do not have a reason to endanger people. Take the necessary steps to open up and extinguish. Be careful.

Contents Fire

Let us move on to the next type of incident we might encounter. The contents fire involves something burning within a building. The greatest concern in these matters involves assessing at what point the fire begins to involve the actual structural components of the building in question (Figure 9.4). Once the fire has grown to involve the actual structural components, the risk and challenge to our forces is increased.

Contents fires can be very hot and smoky. If you are to handle these properly, sufficient suppression water must be moved in via hoselines to the seat of the fire. This must be coordinated with your efforts to vent smoke from the building. Always have your crews check for extension. **An aggressive interior attack that does not kill the whole fire is a failure, and can result in loss of the building.**

Structure Fire

Let us move on to the next logical possibility: involvement of the structure itself. When assessing the situation at a structure fire, you must quickly cover the following items in your mind:

- Type of construction

- Height and breadth of the structure

- Significant dangers, both obvious and hidden

- Help and hindrance factors

- Terrain and access

9.4 Is it a contents fire, or does it involve the actual structural members? *Courtesy of Harvey Eisner.*

You must consider the impact of each factor on your decision-making. Each has a direct bearing on your deployment strategy. You must quickly cover the salient points of each one.

At this point it must be stressed again that the best time to learn which type of structures are in your district is before any incident occurs. It is vital that you go out into your community and learn what is there by making pre-incident planning visits. Being aware of the structural methods and materials used within your district will help you make better decisions when it comes time to deploy your forces. And it is the quality of these decisions that can keep your forces safe.

Type of Construction

Wood Frame Construction. These structures are the ones you will likely see the most often. The more common style, the platform frame, should be built so that there are no openings between floors. The fire should be limited to the floor of origin. But be warned, there are situations where enterprising contractors build around these platforms for their own purposes. Just a couple of years ago we fought a fire in a modern development home. It was allegedly a two-story, platform frame structure. Upon arrival, we noted smoke coming from the basement and both floors of the building. Experience told us that this was a basement fire and that the damage should be limited to that area because of the platform construction we had noted during our visits (We make a point of visiting each of the new developments in our area as they are being built). This gave us an advantage. We knew what lay under the outer covering, or so we thought. However, the fire did not conform to our expectations. A small fire in the basement was quickly extinguished, but fire began to break out on the first and second floors. And it was happening in areas not directly above the cellar fire. In short, it was not acting like it should have. As it turned out, investigators found that the contractors had built wooden cold air return ducts and the fire spread through them. So even if you think you know what is going on, be ready for those surprises that the Fire Gods are preparing for you. They can reach up and bite you when you least expect them.

Many older wood frame buildings feature "balloon-frame" construction. This term comes from the use of an open area that runs within the walls from the lowest to the highest levels of the building. Fires in these types of buildings can cause fire fighting teams to spend quite a bit of time chasing the fire from below. In an older balloon frame structure, you must anticipate where the fire will go and beat it there. These types of fires will require a lot of truck work, such as opening walls and baseboards. You must consider the added staffing requirements for this type of operation.

Ordinary Construction. If the answer to your "What Do I Have?" question involves ordinary construction, be wary. You may be facing a wood frame fire inside a brick oven. And because you have a veneer wall, which may or may not be tied to the structure, the oven can crumble around your ears.

Heavy Timber Construction. As an Incident Commander, you must remember that with this style of construction, you will probably be fighting a "big fire" with "big water." If you have a number of these heavy timber structures in your community, be sure to pre-incident plan for the extra help you will probably need. Mutual aid can be a lifesaver in fires of this type.

Noncombustible Construction. If structural elements perform as designed, you should only have a contents fire in this type of building. However, smoke can still spread by all sorts of internal routes, such as pipe chases, plenums, and utility poke-throughs. As with any other type of fire, you should concern yourself primarily with life safety. This requires a thorough search and rescue operation to ensure that people are not at risk.

Fire-Resistive Construction. Fire-resistive construction is often incorrectly termed "fireproof." This is not so. The truth of the matter is that all of the structural elements are made of materials that are built to stand up to the effects of fire. These elements are also designed to limit the spread of fire.

(**NOTE:** Building construction types and methods are covered in more detail in Chapter 18 of this book.)

High-Rise. Should you encounter a high-rise building when you ask, "What have I got?" be prepared for a different set of circumstances. These will be covered more in depth later in this chapter under Height Considerations, but basically you are dealing with fire on a grand scale. So you must react by beginning to think on a grand scale. In this case you need to say:

"BIG BUILDING = BIG PEOPLE COMMITMENT"

In a high-rise building, it takes more people to perform each task. And you must be able to deploy these forces quickly enough to beat the fire to the punch.

General Building-Related Statements

- If you have smoke on all floors of a structure, you probably have a cellar fire.

- Know the construction types in your fire response district.

- Never trust a truss.

- You must quickly get above a fire in a balloon-frame structure, open up the walls, and halt the spread of the fire.

- Heavy timber buildings will burn for a long time, and then they will fall down.

- Nothing is ever really noncombustible. You just have to get it real hot to see what happens.

- Two buildings on fire = two alarms.

- High-rise fires require large numbers of people who may be exposed to smoke, heat, and fire spreading upward through a number of possible passages.

Wildland Fire

Let us move to our next section of the "What Have I Got?" question. Let us suppose that your answer to this question is a woods or brush fire. This opens up a whole new realm of problems and solutions. And it lays the groundwork for a very fluid situation.

Any fire that starts in the great outdoors has no bounds as to where it can go (Figure 9.5). It can fly like the wind or its sparks can spread for miles on currents of hot air. It responds directly to low humidity and will move as the topography of the locale dictates (Figure 9.6). Of course, fire like this can happen to you in the middle of nowhere, miles from people, water, and mutual aid help. However, it can also occur on the edge of your population areas, in an area known as the wildland/urban interface. It is in these sorts of locations where the potential for property damage, injuries, and deaths

9.6 Topography can play a heavy influence in the spread of fire. *Courtesy of NIFC.*

9.5 A wildland fire can occur at wildland-urban interface. *Courtesy of Mike Wieder.*

can grow. To prepare for this type of fire in your area, you need to gather information ahead of time. Take the time to answer such questions as:

- What are the leading causes of brush fires, woods fires, and wildfires?

- What area of your community is most prone to this problem?

- Can you develop a pre-incident plan for various segments of your community?

- Are there any steps your department can take to prevent these types of fires from occurring?

- Can existing industry and transportation mediums increase our potential for these types of fires?

Answering these questions will give you a good idea of the risk level for this type of fire in your community. The risk could vary from minimal in a downtown urban business district to heavy in a fringe area of a sparsely populated rural community. As with all fires, prevention is the best way to handle a wildfire. With a fire of this type, the suppression effort can be quite labor and equipment intensive. But if you are faced with such a challenge, you must know how to respond to it.

Hazardous Materials Incident

The next type of incident we need to touch upon at this time encompasses those fires and/or incidents that involve hazardous materials (Figure 9.7 a and b). There are three factors that result in different and more severe risks during fire fighting operations. They are:

- Chemical and physical properties of the materials involved

- Large quantities of the materials that may become involved

- Limited special equipment and training

Virtually every fire department has the potential to face a raging gasoline tanker fire. Are you prepared to handle 9,000 gallons of gasoline rolling toward you as a flaming ball of fire? What is your foam capacity? Can you deliver what you have in sufficient volume and quantity? Can you apply it close enough and quickly enough? (Figure 9.8) Are your people trained to employ containment and control equipment to halt the outward movement of flammable product? The answers to these questions are critical. And you had best be asking these questions during staff meetings to plan for future operations!

A critical part of preparing for hazardous materials situations lies in the area of training. All personnel must be trained in the recognition of hazardous materials identification and risk assessment. If people are aware of the potential, they will be better able to determine if a defensive posture will be safer and more preferable. Your pre-incident planning should equip you with a knowledge of the potential for hazardous materials in storage and transit. A good rule could be: "If you don't know, you don't go!"

Really Unusual Incident

The last type of incident listed here is the unusual or uncommon incident. Things such as subway or elevated train incidents, below grade, confined space, silos, or buried vaults can rarely be pre-

9.7a At any time, your department may be called upon to deal with a hazardous material. Be prepared and be careful.

9.7b How about an explosion at a hazardous materials facility? Can your department handle a challenge like this?

9.8 A flammable liquid fire will burn very big and very hot. Suppressing a fire like this will take extra effort.

incident planned to a sufficient degree. You must first decide what should be on your list of uncommon incidents. Then assess what your level of preparation is for such events. After identifying gaps and shortfalls, seek out experts to provide the necessary training. At the very least, learn a lot about how these different scenarios function in a normal way. By knowing as much about normal as possible, you will be better prepared to deal with the unknown.

TIME FACTORS

Let us now move on the arena of time as a factor in your fire fighting operation. Each of these has a direct bearing on your deployment decisions. We look at time in the following dimensions:

Time of day • Time of week • Time of year

Time of Day

What may not be a problem at 2:00 p.m. can turn into a serious matter at 2:00 a.m. At this point I want to caution against the standard obvious comparison of a home in the afternoon, with no one at home, and the same home in the middle of the night. This presumes certain stereotypical patterns that may no longer be valid. It is up to you to determine the demographic patterns in your community. Does a young family with small children live in the same way as a senior citizen couple residing in an adult community? We think not, and so should you. There are as many variations on this thought as there are neighborhoods in your community. You must be aware of how your community lives its life so that you can make intelligent decisions on the fireground.

You must devote a great deal of time to learning about your community. Are any of these situations a problem in your town:

- Is there a rush hour?

- What parts of your town or city are affected by heavy traffic?

- Are there industrial plants? If so, are they on shift schedules? What impact does this have on your response?

- What buildings in your community have a variable occupancy based upon time of day? Know them!

Time of Week

There is a great deal of variability when it comes to time of week as a practical consideration. Let me offer a few questions to stimulate you in this area of thinking:

- Is your middle school going to present the same problems to you on a Sunday that it might on Thursday afternoon at 1300 hours? (Figure 9.9)

- Are you faced with heavy commuting traffic on a daily basis?

9.9 Time of day can make a big difference in the life safety hazards you may encounter in schools.

- Is there a busy time and a quiet time for your shopping district?
- Will your church present the same problem to you on Wednesday that it might on a Sunday morning?

In addition to these basic questions, you need to identify anything that is specific to your community. You cannot learn this information from the comfort of your fire station. You must get out and visit.

Time of Year

Time of year scenarios can vary widely. A fire officer in southern Arizona would surely face different problems in January than we would up in Newark, New Jersey. You need to learn your weather patterns and then gear your operation to the unique problems found in each season. Here are some questions to guide your thinking:

- Do you have a flood season?
- Are there dry desert winds at any time of year?
- Can you move your equipment through the snow-congested streets of your community?
- Can your mutual aid assistance make the run to your town in January with the same speed as in June?
- Are you prepared to fight a fire during a Force 3 hurricane?
- Is your community more heavily populated during certain seasons?
- Is your fire department membership roster steady on an annual basis? Or are there periods when more people leave town on vacation?

This list is by no means complete, but should serve as a starting point to get you thinking. There are many what-if situations that you could face. If you have not spent a good deal of time thinking of different variables and how to deal with them, you will be in for a rough time when fire strikes. Time is a variable in everything you will ever do or face. Recognize it and be ready.

HEIGHT CONSIDERATIONS

How high is up? This simple childhood question takes on epic proportions when it comes to battling a blazing building. It is critically important for you to understand the impact of a building's height on your operational strategy and tactics. So the fourth part of our "What Have I Got?" equation involves the height of the fire building.

In a standard suburban residential development, it may be possible to use ground ladders for any roof-related work you need to perform (Figure 9.10). Usually, however, anything over 30 feet (10 m) in height or with bad terrain approaches requires some form of aerial device. Once again, it comes down to knowing your operational area.

If you have a number of buildings that are three stories or more in height, you will need to consider how to provide aerial service. The standard figure comes from the Insurance Services Office (ISO) Fire Suppression Rating Schedule. Their guidelines state that if you have five or more buildings that are three stories or more in height, or equal 35 feet, an aerial device is needed. If your community does not have an aerial device, you will need to know where the nearest mutual aid aerial ladder is. And then you need to make provisions for its use when needed.

Height can have a direct impact on your ability to rescue endangered persons. In situations involving one- and two-story buildings, you will probably be able to use protective hose streams to reach and assist people who may be trapped. Ground ladders may also prove to be a solution to these types of rescue difficulties. However, as the height of the building grows, so too does the magnitude of your rescue problems.

Height can also have an effect on your ability to perform a number of different fire fighting functions (Figure 9.11). Ventilation, forcible entry, and master stream water applications become more difficult the higher you go. Your early fire fighting decisions in these areas will be influenced by the number of feet above ground level that the fire is burning. Fires above the reach of an aerial ladder must be handled in a very specific way. Access to fire floors can be limited. Water may be a problem. And people will be fighting to use the same stairways to escape that you will be using to go up to attack the fire.

9.10 Is your equipment adequate to assist with your more common structures?

9.11 Fire in a tall building automatically means that you need a lot of people.

Structural height is an important component of your fireground operation. Know what your options can be and drill on them ahead of time. Then, when the time comes, react to the situation in a calm and studied manner.

SUMMARY

In this chapter we have covered the basics of how to answer the first question in our fireground command formula: "What have I got?" The components of that question break down into four parts:

1. What type of **W**eather-related problems do I have?

2. What type of **I**ncident are we facing?

3. What are my **T**ime-related problems?

4. What are my **H**eight-related problems?

If you can quickly answer your WITH questions, you can then move on to attack the fire. You proceed at your own peril if any of these questions goes unanswered.

10

Two Key Questions: Where Is It and Where Is It Going? How Can You Fight It If You Cannot Find It?

The last chapter dealt with your first sizeup question: "What Have I Got?" Knowing what you are facing is the first step. But how are you going to get the job done if you do not know where the fire or incident is? The purpose of this chapter is to discuss a number of ways in which you can find your enemy – fire — and then work to limit its travel. We have decided to combine questions two and three — Where is it? and Where is it going? — because a fire's location and its potential for travel are tied closely together.

There are those in the fire fighting business who have told me of a "sixth" sense possessed by certain Incident Commanders. These are the people who seem to know where a fire is through some seemingly mystical power. A closer look reveals that these individuals also have a great deal of experience and practice the basic principles of effective fire fighting. To understand how you can discover just where a fire is, you must first master the five senses possessed by all human beings:

- Sight
- Hearing
- Smell
- Touch
- Taste

SIGHT

Let us look at sight as a means for discovering where the fire is located. As you roll out of the fire station, you know that you are responding to 101 South 10th Street. You know where this address is located. While responding, you note a dark, puffy plume of smoke, so you have a strong clue that you are about to face a working fire. As you round the corner into South 10th Street, you are greeted by a most informative sight. You see a 2-story residential home with flames blowing out the second-floor windows. You now know where the fire is, and you know that is on the second floor. You also can begin to ponder where the fire might be going.

When a structure is fully vented, you can usually look at the flames and tell where the fire is. Unfortunately, it is not always this simple. What you see might just be the tip of the iceberg. But it sure is nice to have a tip to start the trip. If you just see smoke, the smoke will tell you which building has a fire in it. But you will have to remove the smoke from the building to identify the actual location of the fire (Figure 10.1).

There are also instances where you must actively hunt for the seat of the fire. During times like this, your eyes can give you a number of clues as to the fire's location. Some of the things you must look for inside a structure are:

- Discolored paint on the walls

- Discolored, dried, or peeling wallpaper

- Puffs of smoke from behind door frames and window sashes

- Discoloration around electrical outlets, utility locations, cable television openings in the walls, or the points at which plumbing pipes or other piping pass through the walls, floors, and ceilings

- Smoke puffing through cracks in the walls or between the baseboards, floors, and walls

10.1 Frequently, you will have to allow the smoke to escape before you can see where the fire is. *Courtesy of Mike Wieder.*

Recent advancements in technology have made several types of electronic devices available to us. These appliances can locate heat sources in environments that obscure them from the naked eye. One such device can sense temperature differences of as little as 0.5 degree. These devices, while quite expensive, can repay their purchase price several times over in reduced property damage and human lives not lost to fire.

SOUND

How about sound? Is it possible that your hearing can help you find a fire? Absolutely; it has helped me many times. As you are advancing a charged hose line into a smoke-filled environment, you may wish to pause and listen for the sound of burning. In the standard wood-burning scenario, there is a discernible sound. If you can pinpoint the sound, you can move in the direction of the fire.

SMELL

How about smell? As you gain experience in fire fighting, you find that many materials burn with a distinctive aroma. In the days before SCBA, this was a way of identifying what was burning, and which part of the building might be the source of a particular aroma. See if you can identify, in your mind, the smell associated with each of the following:

- Unattended cooking

- Oil burner delayed ignition

- Ballast burning in a fluorescent light fixture

- A fire in the woods

- A car fire

These can still be clues, but remember that everyone must be a great deal more careful now about exposing their lungs to the products of combustion. Do not expose yourself to hazardous vapors just to find out what is burning. Use your self-contained breathing apparatus.

Let me share a little story with you. It is about how a potentially devastating fire was averted by a firefighter's sense of smell. Or at least the sense of smell, as augmented by a great deal of patience and diligence. I was the Captain of an aerial ladder company. Our unit was responding to a telephone report of a fire in a small neighborhood grocery store. Upon arrival we were met by the owner. We entered his store and encountered a very slight aroma of burning wood. After a preliminary search of the structure, we cut back on the alarm response and continued our search of the building. We fought the urge to give it a quick look and an early departure. Our units checked the basement, the first floor store, and the second-floor apartments, and still we found nothing.

In order to clear our sense of smell, we went outside for a break from our search. We discussed our problem at length. After we went back into the store, we were able to trace the aroma to a display window near the front entrance. As part of our discussions, we discovered that some building alterations had just been completed. We decided to do a little bit of opening up to see if we could find a construction deficiency. It was most fortunate that we stayed around to check. After removing a window sill, we discovered a burning area in the wall. It seems that a single nail had been driven through an old, dry electrical wire. The resulting short in the wire had caused it to overheat. Eventually the wood underwent pyrolytic decomposition. If we had just given it a quick look and left, we would have returned before too many more hours had passed. The smell tipped us off. The smell was the clue.

TASTE AND TOUCH

The last two senses – taste and touch – should be used as infrequently as possible on the fireground; nonetheless, there are some occasions when touch can be a deciding factor in finding out where a fire is. Using a gloved hand to feel for heat in a wall is an old method of detecting hidden fire, but still a good one. By using the back of your hand, you minimize the chances of injury because your muscles will contract and pull your hand away from a hot wall (Figure 10.2).

A word of warning at this point: The fire will really have to be in a hidden location for you to use this method as an initial means of locating a fire. Nonetheless, knowing how to

10.2 Use your sense of touch carefully, but the back of a gloved hand can provide useful information.

handle such a situation may prove invaluable. Remember that in order to beat fire at its own game, you must use all of the skills at your disposal.

Each of our senses can provide vital clues. They can help you pinpoint the location of the fire by what you can see. Once you have located the fire, you can begin to decide where it might be headed. As unpredictable as fire seems to be, it responds to the laws of chemistry and physics, as well as to the physical forces of nature. By knowing the environment in which fire lives and works, we can begin to develop some clues about how to predict its spread. Let us look at the nature of these forces and their effect on the spread of fire.

HEAT TRANSFER

How does heat travel from one substance to another? Let us offer an example that will clarify this matter. If you think of two things touching, it is very easy to visualize how something might move from hotter to colder. Look at what happens when you place a pat of margarine on a piece of toast. The margarine is heated and melts. The toast is cooled and accepts the melted margarine into the porous areas of its surface.

Let us now translate this into our fireground scenario. First and foremost in your bag of analytical tricks, you must remember that heat rises. It does this because the heat of the fire is seeking to be calmed by the coolness of the surrounding air. And it will go to great lengths to reach up to that cooler environment. That would mean that any basement fire that you encounter is seeking cooler air on the first, second, and third floors. And the more floors you have above a fire, the more places that fire will want to go. This is a rule of thumb that you can take to the bank. As it rises, smoke also poses a life hazard to the areas above a fire.

As was stated in the chapter on fire behavior, a fuel's position, manner, and surface area play a great part in determining a fire's progression. A flammable or combustible material in a confined area, without a great deal of air movement, will not burn as quickly as the same material exposed to outside air currents. A fire's movement is governed by the standard heat transfer mechanisms (Figure 10.3). Fire will travel via:

- Conduction
- Convection
- Radiation
- Direct flame contact

It is up to you, as an Incident Commander, to apply these laws to the situations you encounter. You must never forget that these are constants in your life.

It is during this initial phase of the "Where is it going?" scenario that another question should pop up from your mental bag of tricks (Figure 10.4). Ask yourself, "Can something bad happen from the picture that I see in front of my eyes?" And that something bad can take a number of forms, such as:

- Will the building blow up?
- Might the building fall down? (Figure 10.5)
- Can we limit the fire to one area of the building?
- Can we limit the fire to one building?

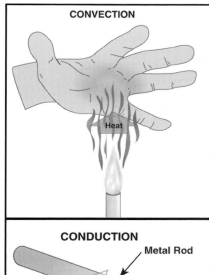

CONVECTION

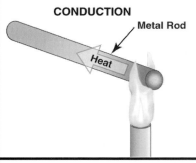

CONDUCTION

Metal Rod

Heat

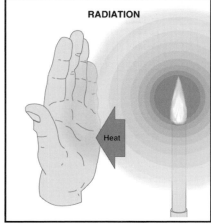

RADIATION

Heat

10.3 When you are asking where the fire is going, remember how it goes: by convection, conduction, or radiation, and assess the risks you see accordingly.

10.4 "Where is it going?" — right to the exposures if you are not careful. *Courtesy of Ron Jeffers.*

10.5 The fire might also be going down, as in the structure posing a real collapse hazard. Be careful. *Courtesy of Harvey Eisner.*

- Are there buildings nearby that the fire can spread to?
- Can we limit the fire to the community of origin?

There are also some ins and outs of the construction world that we need to introduce at this point. There are a number of ways fire can travel through the openings of a structure:

- Within walls in balloon frames
- Unprotected vertical openings

- Via beam ends in construction with metal structural members
- Upward through shafts (elevators, dumbwaiters in older buildings, pipe and utility chases
- Upward through the central core of high-rise buildings
- Laterally through the air-handling duct system
- Behind wooden facades
- Through hidden areas that have been enclosed during renovations
- Via windblown embers from a burning wooden building

It is important to remember that the size of a fire will have a great bearing on its potential for spread. There are simple reasons for this:

- A big fire will burn over a larger relative fire front
- The heat source will be larger
- While the heat given off from a small fire is the same as from a larger fire, there is just a lot more of it: that means a bigger fire will probably cause you more problems than a small fire

SUMMARY

This chapter has been designed to arm you with the facts that you must have to conduct an initial fire evaluation. Once you have determined that you are facing a fire, you must quickly answer two important questions:

1. Where is it?

2. Where is it going?

Once you have made these assessments, you can then move on to resource allocation decisions. These will be covered in the next chapter.

11

The Fire: We Have Found It... Now What Have We Got To Stop It?

Picture yourself in front of a burning building (Figure 11.1). It is a two-story wood frame structure. There are no nearby structures. The homeowner is standing next to you with his wife and children. The dog is safe too. Now what?

It is at this point that you will enter the world of resource deployment. Here is where you have to answer the question: "What resources do I have to attack and extinguish this fire?" There is also a corollary question that must be answered immediately after this one: "Can they get the job done safely or are more resources needed?" As the Incident Commander, you will have little more than a radio to extinguish the blaze in front of your eyes. You must first assess just how much in the way of people, pumpers, and water you will need. You must then do the following very quickly:

11.1 Now that you are sure what you are facing, what have you got to stop it? *Courtesy of Joel Woods, University of Maryland Fire and Rescue Institute.*

1. Deploy your response force as per your operating guidelines.

2. Call for as much help as you feel you need. Do not be conservative. You can always return units you do not need.

3. Establish a reserve requirement/equipment pool.

4. Call for the necessary resources to staff both the attack and reserve forces.

5. Pay attention to the operation as it unfolds before your eyes.

Let us remember one fact that is frequently overlooked. It is important to have a plan for conducting fireground operations before they happen. This allows you to have a framework within which to develop your decisions, and you will know that everyone is operating from the same sheet of music.

GENERAL OPERATING GUIDELINES

Because of this fact, a few words need to be devoted to developing your fire department's deployment strategy. It often appears to the unschooled outsider that we perform a series of disjointed operations that happen to extinguish fires. They believe that we are "winging it." Do not let yourself be deluded into this idea. We must have a definite deployment plan if we are to do our job correctly. To do this you should develop a series of General Operating Guidelines (GOG's).

Your guidelines should define what fire companies are expected to do, not how they are to do it. This is an important distinction in well-written GOG's. As a part of these guidelines, your department should indicate the tasks, duties, and responsibilities of the various units that respond on an alarm. The tasks should be loosely defined, so that units are not trapped into doing the wrong things. You want to offer guidance, not tie people's hands. In order to develop well-written guidelines, you must determine how many fire department resources you need to safely protect your community. It is not this book's task to teach municipal master planning. However, that same methodology can form the basis for your deployment strategy.

An excellent source for planning department resources comes from the National Fire Academy. Much of what follows is extracted from their programs. A framework has been developed that allows for the development of deployment strategies. These strategies are developed to suit the needs of a community. And they are developed from information that is generated locally. In that way they are tied to the identified needs of a particular community.

COMMUNITY RISK ANALYSIS

Your first task in a community study deals with a consideration of risk. You need to break your community down into manageable bites in order to analyze the demands created by differing areas. By examining such elements as occupancy hazards, demographic patterns, and past fire history, an overview of emergency response requirements potential can be developed (Figure 11.2 a - c).

Once the community's needs have been identified, you can move to the next task, which is to assess the delivery capabilities of the fire department. This allows you to address any gaps, shortfalls, or excess capacity in the delivery of fire fighting services (Figure 11.3 a and b).

The last stage of your preparation to combat fire in your community is critical. It comes when you create a fire department that can handle the risks it faces. By working your way through this planning process, you will become intimately familiar with how your department works: You will know what you have, and you will know what it is capable of doing. And you will know when something is beyond your ability to control.

EMERGENCY SCENE ASSESSMENT

It is equally important that Incident Commanders be able to look at a fire scenario as it unfolds. They must be able to assess the fire's development. They must also be able to quickly judge the potential requirements for apparatus and personnel. There is a wealth of information that exists which, if you review it, can assist you in determining the resource needs for a fire fighting operation.

These fireground command skills come with experience and training. You must learn to balance the resources you have against the demands of the situation. And you must be flexible as things play

11.2a The differing parts of your community will demand different resources during an emergency.

11.2b The variety and size of structures in a crowded urban area is considerable.

11.2c Know the occupancy types and life hazards in your response area.

11.3a You must have apparatus that is equal to the task at hand. *Courtesy of Harvey Eisner.*

11.3b And it must be stationed so that it can be used for maximum effectiveness.

out during the fire fighting attack. The chapters that follow will detail a range of deployment options. However, you are still the one who must make the critical decisions under fire. Once you have decided the level of resources needed in a particular situation, another contingency must be addressed. You must maintain an on-scene reserve force to handle sudden, additional needs.

Experience has taught me that it is always better to call for help before you actually need it. The military does this when it attacks an enemy objective. They may choose to press the attack on a small scale. In this case they will conduct the assault with two platoons of infantry, while keeping a third platoon in reserve to handle contingency matters. They might also augment the force with a special weapons squad, with heavy machine gun and mortar teams.

DETERMINING NEEDED RESOURCES

The same parallel holds true in fire fighting. For example, suppose that you decide you can handle a structural fire attack with two engine companies and a truck company. To provide an additional margin of protection, it would be good to summon an additional engine and truck company for standby. If needed, they are available. If not, they can return to service. By doing this, you will be ready for any unforeseen circumstances. An Incident Commander can never call for additional assistance too quickly. There are times when Incident Commanders have analyzed a burning building into a smoldering pile of rubble. No decisions were made, but a lot of decisions were pondered.

Rapid Intervention Teams

For larger-scale fire attack operations or unusually risky incidents, a specific company or companies should be assigned duty as a rapid intervention team. Over the course of the past decade, a number of firefighter deaths or serious injuries led forward-thinking fire officers to decide that a special unit was needed at the scene of fire fighting operations to rescue teams of firefighters who might suddenly become trapped or exposed to a dangerous situation. These teams should have the necessary tools and talent to save lives. The Firefighter Safety and Assistance Team (FAST), is another name for this concept. I know of fires where lives have been saved because someone was there with the tools and talent to make the necessary rescue.

Getting the Most From Your Resources

How do you learn to deploy your forces and summon reserves and added assistance? The best way is through experience, supplemented by training. Fire officers can definitely improve their learning curve through the use of structured training. What else can you do? Read about fire fighting. Train in conducting fire fighting operations. Hold drills and respond to actual fires (Figure 11.4 a - c). Then conduct critiques and apply the lessons learned to your training programs. You will become more comfortable with assessing what a fire will demand from your attack forces. By using the combination of techniques listed above, you will become skilled in delivering the forces at your disposal in an effective manner. More important, you will know when to call for help. And you will do it in a rapid-fire manner.

You must begin to think numerically when looking at a fire building. Do not limit your view to smoke and flames. Look for numerical needs. For example, in addition to thinking about construction, volume and location of fire, you must begin to think of attack force size.

You must also learn to think about estimating structural sizes. You must look at a building and determine its relative size. Once you have done this, you can begin to arrange your personnel and apparatus response based upon your needs assessment. In this way, you can then think in terms of force deployment in terms that mate up with the fire structure you are about to attack. In this way your mind allows you to create the resource estimates that form your decision base. You can then request and deploy an adequate force to stop the fire in its tracks. Here are some of the things you need to think about as you begin your fire attack:

- The minimum response recommended by the **NFPA Fire Protection Handbook** is 12 firefighters, two pumpers, one truck company (or a unit capable of truck company work) under the command of a chief officer.

11.4a Your training program must be ongoing. *Courtesy of Ron Jeffers.*

11.4b Always emphasize safety – always. *Courtesy of Ron Jeffers.*

11.4c When you face the real thing, proper training will enable you to get the job done efficiently and safely. *Courtesy of Ron Jeffers.*

- Most fires are extinguished by a single hoseline.

- Always back up your first hoseline with one of equal or larger size. Each line needs a minimum of two people to advance and operate it.

- A crew of two people to handle utilities control functions (light, gas, water).

- Two people per 2,000 square feet for search and rescue.

- Minimum crew of two for roof ventilation where needed.

- Minimum crew of two for forcible entry and interior truck company work.

- Think of two attack lines per floor as an operational minimum for standard residential scenarios. This applies to detached, single-family dwellings. You will need to think of larger fire areas in similar terms.

Take a look at the fire in this picture (Figure 11.5). Will a minimum response, as specified above, be sufficient to locate, control and extinguish this fire? If not, how many more will you need? Where will you get them? Now take a look at this fire building: (Figure 11.6)

11.5 How many people will you need to safely handle a situation like this? *Courtesy of Harvey Eisner.*

11.6 A fire of this size will require many more people to attack safely. *Courtesy of Harvey Eisner.*

Will a minimum response, as specified above, be sufficient to locate, control and extinguish this fire? If not, how many more will you need? Where will you get them? It is this type of analysis that must be done during the heat of battle. A modern Incident Commander must master this approach to operational thinking. This format serves as the basis for our approach to fireground command and control.

The formula is really quite simple: it requires that you commit a series of questions to memory. You task is then to ask these questions during the stress of a fire scenario. The answers you formulate will serve as the basis for your operational decisions. In this way, you can approach every fire with a common thought process. Although each fire will have certain differences, your ability to sort through data will speed up your ability to spot the site-specific differences. By identifying the similarities quickly, you will be able to identify and attack the site-specific elements more quickly.

SUMMARY

In this chapter we have taken a basic look at resource deployment. Quite simply, we have offered another question that must be answered at every fire you will ever face. "What resources do I have to attack and extinguish this fire?" A second question that must follow very quickly on the heels of the first is this: "Can the people and equipment who are en route get the job done, or are more resources needed?"

You will need to learn a few facts in order to be able to answer these questions quickly and effectively. You must discover what your community may require of your fire department. And you must develop a thorough understanding of what your department can deliver. Knowing what might be needed and what can be accomplished will allow you to develop a more effective fire department.

It is your knowledge of all of the above factors that will stand you in good stead when the time comes to do battle. An intimate familiarity with resources and their deployment will equip you to do battle effectively. Remember that your skill as an Incident Commander comes from a mix of learning, training, experience and evaluation. Leave no part of this equation out of your fire service career development program.

12

The What Do I Do and Where Do I Get Help? Phases of the Operation

A great deal of time has been spent bring you to this point in our text. It is in this chapter that we will provide you with some practical examples of actions you might find appropriate for attacking and extinguishing a structural fire.

It is at this point where the rubber begins to meet the road. This is the moment when you as the Incident Commander become responsible for placing your personnel in harm's way. This chapter will outline a series of options for your review and give you guidance as to when to use each attack option.

At the very lowest common denominator, you have but a single decision. You can choose either to:

- Fight the fire

- Not fight the fire

12.1 Fire in a structure like this signals extreme caution: the collapse potential is severe. *Courtesy of Harvey Eisner.*

If you think about it, this theory is similar to the classic view of the way adrenaline works in the human body. Theorists tell us that adrenaline stimulates us in our fight or flight decisions. If we need to get away, it stimulates our body to run faster. If we need to fight, it allows us to draw upon our deepest reserves of strength. This is, in a sense, how we approach fires and fire fighting. While it might be hard to imagine a situation where firefighters would run from a structural fire, such times exist (Figure 12.1). How about situations involving explosives, hazardous materials, or biological agents? The presence of hazards like this should cause us to pause or pull out. What about the potential for a disastrous structural failure? We would sure want to flee such a scenario.

Firefighters frequently have to flee wildland fires (Figure 12.2). I can remember an incident where a water drop from a small biplane saved our collective butts from a fire that was leaping across the tops of a stand of pine trees.

It is not easy to compile a complete list of when to fight and when to flee, but here is a simple guide for making this decision: If your sizeup tells you that your people could be in danger, be careful. Whenever possible, avoid placing them in danger. This is not an endorsement of cowardice. It is a reminder to take care of your most precious resource: your people. The risk you expect your people to take must be at least equal to the benefit you expect to gain. Your goal, of course, should be to return your troops home safely after every incident. Remember your job, but don't forget your people.

Let us now move along. By this time in our textbook of fire fighting methodology, (meaning, by the time we proceed to ask ourselves what should we do) we have arrived at a point where we have assessed the potential for exposing our people to danger. When you have decided that you do not have to flee the fire, you must then decide which steps to take to control the fire you are facing. As stated in an earlier chapter, the suppression options available to you are as follows:

- Aggressive interior attack (The Cavalry Charge)
- The "Blitz Attack" method
- The "Do a Blitz Attack and think about moving in" method
- The "keep it to the block of origin" operation

The choice of initial attack option is very important. You must weigh how much fire you face against how much fire department attack capability is available to fight it. What follows are some examples of each for you to add to your personal data bank. Remember to memorize these types of experiential advice. In all my years in fire fighting, I have never been blessed with the ability to set up a podium in front of a burning building and work from my notes. Because you are not likely to be so lucky, either, you will have to work from your own mental bank of facts.

Take a look at the picture on the next page (Figure 12.3). What resources do you think are needed to quickly attack this fire? I would suggest that an initial attack hoseline (1¾-inch or 2-inch), backed up by a similar line, would deliver sufficient flow to attack and extinguish this fire. Those hoselines would have to be moved in quickly and directly to the heart of the fire. This task is likely to require four people to operate and move the hoselines. It is my firm belief that that the minimum number of hoselines for a safe and effective fire attack is two: primary and backup.

As a fire grows in size and scope, additional multiples of two should be deployed. As you can well imagine, big fires require a lot of people to handle safely. You must constantly consider whether you have enough people to get all the necessary jobs done. Always be thinking and anticipating what the fire might do and where it might go. These two questions, which were covered in earlier chapters, need to be asked periodically to keep you thinking. In this way you may be able to outwit the wily old fox we call fire.

AGGRESSIVE INTERIOR ATTACK

Let us take a look at what experience tells me constitutes a minimum structural response assignment. At the very least, you will need an attack pumper and a supply pumper. This allows you a primary capability and a backup in case of pumper or water supply problems. You would need two drivers to operate these pumpers. Two firefighters will be needed to perform ventilation. An aerial ladder or service vehicle must be available to carry the necessary tools. Two more firefighters are

12.2 Be sure you know the hazards of working in wildland fires. *Courtesy of California Office of Emergency Services.*

12.3 Have you stretched enough hose to do the job safely? *Courtesy of Mike Wieder.*

necessary for the delivery of search and rescue capabilities. You should have one or two more firefighters to cover utility service cutoffs. To be safe, you should also have an interior sector commander to assist you as the incident commander. When possible, assign a company of at least three (3) people to function as a safety sector. The officer can perform the function of incident safety officer, and the other two personnel can be ready to assist should trouble arise.

So what are your basic requirements for implementing an aggressive interior attack?

- Two pumpers
- One aerial ladder or service support company
- One incident commander
- One interior sector commander
- Two pump operators
- Four people operating hoselines
- Four people on ventilation and search and rescue
- One or two on utility control
- A company assigned as a safety sector team

In order to mount an aggressive interior attack, you will need a force of two engine companies, one truck or service company, a company to provide safety services, **and 16 or 17 people**. This should allow you to complete all tasks in a safe manner. Has an operation of this nature ever been done with less? Probably, in certain isolated areas. But not safely. Experience has taught us all a very hard lesson: Fire fighting is an extremely dangerous and labor-intensive undertaking.

Let me say it in plain language: **If, as Incident Commander, you ask too few people to perform too many tasks, you risk killing or injuring them**. No matter what your individual feelings about mutual aid are, you must reach out to your neighbors for help (Figure 12.4). You must then deploy your help as it arrives.

This "What do I do?" phase of your fireground operation ties in directly to the "Where do I get help?" phase of your fireground operation. This is critical, because if your assessment tells you that you need 16 people and you can only count on having 10, you have a problem. Finding those people requires preparation. You need to know who to call for help well before the first flames blow out the window. Better yet, your pre-incident planning efforts should generate automatic aid responses. By arranging for automatic help to be en route, your decision time frame is shortened. And if it turns out that you do not need the help, just send them home via a simple radio message.

As was noted before, resource availability serves as the basis for a number of the questions you will ask during the "What do I do?" phase. My goal remains simple: to get you thinking in numerical terms. It is very important to think in terms of pumpers, hose, ladders, and people. But it is even more important to think of how many of each will be needed to do the fire fighting job you face in a safe and efficient manner. If you do not have sufficient immediate attack forces, what are your alternatives? I would suggest that at this point you begin weighing the other attack options previously listed above.

I know that many of you will be unable to generate the numbers listed in this text. I also know that you will do what you think is right to get the job done. Remember, you should not even entertain the idea of violating the Occupational Safety and Health Administration's (OSHA) respiratory protection standard. And be ready with your plan to meet the two-in/two-out provisions of this mandate.

You will have to be guided by your view of the fire as it unfolds in front of you. You will then have to modify this view by using your knowledge of the people involved in the operation. It is important that you have a good working knowledge of what your people, as well as the people from your mutual aid assignment, can be expected to accomplish. Think about who is coming (Figure 12.5). Who are they (hometown or mutual aid)? What is their training level? Will they bring the necessary equipment? Are they rookies or veterans? How many of each are in the force structure on scene? How have they performed for you in the past? A good Incident Commander will take great pains to learn all about his or her people. You must also know about the people who make up your mutual aid network. Who are they? How good are they? And how quickly can they arrive? These are event-specific questions that you must be prepared to answer, and how well you do so can determine the success or failure of your operation. And how well you perform pre-incident planning will determine, in a great many instances, how well you succeed on the fireground.

BLITZ ATTACK/THINK ABOUT MOVING IN

A good way to buy yourself some time is to consider opening up with a blitz attack (Figure 12.6). Although you do want to buy some time, you do not want to let the fire get completely out of hand. You do this by first applying big water from the exterior. You do this while other members of your team are stretching an interior attack line. It could be that enough fire is knocked down by your use of a pre-piped master stream device to allow for the rapid entry of an interior hoseline attack team. As time permits, a water supply source must also be established.

There are a couple of critical points that have to be addressed here:

12.4 The Incident Commander must communicate frequently and clearly with all units. *Courtesy of Ron Jeffers.*

12.5 It is always best to call for help soon. If you find that it is not needed, the extra units can always be turned back.

- Are building occupants in danger inside the building?

- Will your use of an outside master stream endanger them? Are these people trapped?

- Will they die if you do nothing to knock down the fire and move into the structure?

To help you with this, let me offer a great principle to guide you in making this type of life-and-death decision. In just about everything I have ever read or studied about fire fighting, I have encountered words of guidance in this matter, and these words have been at the root of my thought processes throughout my fire fighting career: **You can save a lot more lives if you put the fire out before it kills the people.** Simple, but true.

12.6 Using a blitz attack can knock down enough fire to make an interior attack possible. *Courtesy of Ron Jeffers.*

A corollary to this comes from the late Emanuel Fried. He often said that the best way to make a rescue was to put a hoseline between the fire and the people in danger. During the course of my career, I have seen a lot of lives saved by a single hoseline. In these cases, a stream of water was placed between the fire and the people who were trying to get out of a burning building. It is not as spectacular as carrying people down ladders, but it is a lot more efficient.

Some experts have suggested that we should do nothing at the scene of a fire until our resources are equal to the task at hand. I do not subscribe to this. Other veteran fire officers have suggested that we should mount stacks of firefighters' bodies on the fire until we smother it. Others tell us that we should attack fires with the fierceness of kamikaze pilots, in this way saving countless people. I do not believe in these other approaches, either.

In those cases where resources are not immediately equal to the task at hand, steps exist that can be undertaken to begin saving lives and property. They might not be as effective as a direct interior attack, but they can have a positive impact. They can buy time until the cavalry rides into view. It is these steps that make up what we call the "blitz attack." They are:

1. Attempt to determine the degree of life risk.

2. Use a pre-piped/pre-connected master stream to darken down the fire.

3. As soon as possible, stretch a supply hoseline to a water source. (Read that tankers if you are rural in nature)

4. Move an attack handline into position and await sufficient forces to put it into service.

5. Radio incoming units of your plan.

6. As these units arrive, deploy them into the standard operating positions for an interior attack.

7. If and when the time and talent is right, move in and attack the fire.

By operating in this manner, you can begin to do something about the fire and simultaneously begin to deploy your forces for an interior attack operation. If you have to mount a strictly defensive attack you have already begun; if adequate attack forces arrive, you will have already begun to clear the way for an interior attack (Figure 12.7).

Coincidentally, these are the same steps you would take for the development of a "blitz it and think about moving in" type of fireground attack. You will begin with the exterior attack steps and think about what might be done if you see that your master stream attack has a chance of succeeding. In this case, it is a matter of degree. How much fire are you facing? How large a force can you quickly deploy? And how soon will the help you have called arrive? My department has experienced some great successes by kicking the fire in the teeth and getting ready for an interior attack.

HOLD IT TO THE BLOCK OF ORIGIN (THE DEFENSIVE MODE)

If all else fails, you still have one more option. You can resort to what we call the "hold it to the block of origin" operation. Others have chosen to call this the defensive mode. However, my term reflects what I have seen and done over the past three decades. Sometimes you should consider yourself very lucky if you can actually limit the fire to the block or area where it started (Figure 12.8). When confronted with a fire of this magnitude, a couple of thoughts should strike your mind:

• This seems like more than we can handle.

• No one should be sent into that building (or group of buildings).

• Can we stop this?

I can remember one fire in particular where these questions came quickly into my mind's eye. It was on a cold, dark, and windy November night. As we rolled in on the response, we could see embers blowing up into the night sky from blocks away. Prior to our arrival, the first-due engine company called for a second alarm. From five blocks away we were greeted with the view of a wall of fire looming over the surrounding tenements, factories, and single-family homes. As I parked my vehicle and stepped out to confer with the Deputy Chief, the thought occurred to me that we might not be able to stop this fire. We were faced with the following factors:

- A 60-knot wind
- A large two-story, brick warehouse fully involved in fire
- Three adjoining factory buildings — 2 four-story brick and one five-story brick
- A heavy ember condition due to the wooden roof being well involved and exposed to the wind
- A tightly congested, urban residential, wood-frame housing neighborhood about a block away, downwind from the fire
- City-wide deployment constraints owing to an earlier warehouse fire
- Tired firefighters who had been at two warehouse fires during the previous night's tour of duty

  After a very short conference, I joined the Deputy Chief in deciding the following:

- This was to be a defensive battle.
- We were going to attempt to keep this fire from leaving the block of origin.
- We decided that our resources needed to be augmented and a reserve force established. To do this, a third alarm was sounded.
- Given the wind and ember conditions, we decided to request two additional engine companies for ember patrol on the east or downwind side of the fire.
- To pray a lot.

12.8 If a fire is big enough and advanced enough, holding it to the area of origin will be considered a success. *Courtesy of Harvey Eisner.*

12.7 Even after a great deal of water is applied, the situation must be reevaluated as to whether it will become an offensive or defensive attack. *Courtesy of Harvey Eisner.*

From this point on, it was a matter of pouring water on the main body of fire, as well as on the exposures. Fortunately, water supply was not a problem in this industrial area. We had the benefit of both high- and low-pressure water systems. Not too far into the operation, units operating to the south side of the fire identified sprinkler systems that were operating in two of the adjoining factories. It was decided that these systems had to be supported. Using these in-building systems played a crucial role in limiting fire spread on that side of the fire. After an all-night battle, this fire was placed under control. Fire units remained on location for many hours as the building was demolished.

While the building was lost, our strategy prevailed. The surrounding buildings were saved. No other fires resulted from downwind embers. The surrounding buildings suffered only moderate damage on a limited number of floors. There were no injuries to the more than 60 fire people who operated at this fire. This operation was deemed a success.

This incident proved to be a true test of our decision-making abilities. I suggest that you will be well served by working hard to get better at making these types of decisions. Remember, proficiency comes with experience and exposure to fire fighting operations, and is supplemented by a lot of reading and reviewing. Of course, you will also need a conscious desire to get better at what you do. You must conscientiously review each fire you attend. You will need to discuss it with your superiors, your peers, and your subordinates. Never overlook any source for making your decisions better. Compare these fires to what you have read about in the books. And think about how you want to do things differently in the future.

SUMMARY

This chapter discussed the "What Do I Do?" and "How Do I Get Help?" aspects of fire fighting. Our first response led to another question: Do we fight or do we flee? Rest assured that there are circumstances that would require you to flee. Disasters involving the forces of nature head our list. Hazardous chemical involvement would be another entry. A raging wildfire qualifies as a good reason to retreat. And never forget that dangerous structural conditions normally indicate that you adopt a no-fight posture.

More often than not, you will be required to take a positive fire fighting action. In that case, you need to be mindful of the four types of action available in your fire fighting arsenal. As stated earlier, they are:

- Aggressive interior attack (The Cavalry Charge)

- The "blitz attack and move in" method

- The "blitz attack and think about moving in" method

- The "keep it to the block of origin" operation

It is critical to master the skills necessary to operate in each of these modes. This will require the use of drill ground preparation that ranges from the very simple to the very complex. At one end of the spectrum you will be laying individual hoselines, raising ladders, and operating pumpers. You must then combine the acts, actions, and activities of individual units into cohesive teams. And for the largest fires you must be able to deploy groups of teams against a major scenario. These types of command and control skills only come with training and experience. So get started now.

13

How Am I Doing?: The Question of Accomplishment

By the time you reach this chapter, you should have a good idea of how you would handle a working fire. The intent of the previous chapters was to help you learn to assess and act in a step-by-step manner. The last chapter covered the options available to you in the "What do I do?" and "Where do I get help?" phases of fireground operation.

Resource availability served as the basis for a number of the questions asked during the course of that chapter. My goal is to get you thinking in numerical terms when you are assigning pumpers, hose, ladders, and people:

- How many people?

- How many pumpers?

- How much other stuff do you need to get the job done? (Figure 13.1)

- Where can you find these things?

- How should they be applied to the problem at hand?

- How many of each will be needed to do the job in a safe and efficient manner?

That last point serves as the topic for this chapter: showing you how to assess your potential for success. How will you know that you are succeeding if you have no idea what success should look like? The whole point of this premise is that you should be aware of the fact that you will need to set goals for yourself. It is by setting and attaining goals that one becomes a better participant in any endeavor.

Just what is a goal? Goals can best be described as ambitions we wish to achieve, or things we want

13.1 Learn to think in numerical terms when assigning people, pumpers, and so on. *Courtesy of Harvey Eisner.*

to do. What would be a good example of a goal, taken in the context of fireground operations? An example might read something like this:

**"It is the goal of our fire department to limit 95 percent
of all structural fires to the room or floor of origin."**

This is something to which you can direct your efforts and the efforts of your people. Let us imagine that you arrive at a working fire and wish to work toward this goal. What will you do? You begin by using the first question in your arsenal of strategy and tactics: "What have I got?" Apply that to the fire you see here (Figure 13.2).

Your answer: A working fire in a structure.

You ask your next question: "Where is it?" When you apply that to the fire above, your answer is fairly simple: In the building. At this point, how do you apply your goal of limiting 95 percent of your structural fire incidents to the room or floor of origin? You will do this through the mechanism of an aggressive interior attack on the fire using at least two attack hoselines.

Now ask and answer your next question: "Where is the fire going?" It appears to be about to envelop the first floor and possibly spread to the second floor. If it does this, will you be able to meet your goal of limiting the fire to the room or floor of origin? The answer is obvious: NO. So you must do something to see that this spread does not occur.

Here is where your next question can begin to tell you about your potential for success: "What have I got to stop it?" Remember that you said just a few sentences ago that you would need to use at least two initial attack lines to mount an aggressive interior attack on this particular fire. Do you have those resources en route in a timely fashion? A yes answer is good; it allows you to pursue your goal with a chance for success (Figure 13.3). A no answer will force you to select an alternate means of reaching your goal; or in a worse case, selecting a different goal.

13.2 How will you limit this fire to the area of origin? *Courtesy of Bob Esposito.*

13.3 Proper resources, training, and incident management set the stage for a successful outcome.

The yes answer leads you in a positive direction toward the answer to your next sizeup question: "What do I do?" This is a simple question with some very strong consequences. Simply stated, if you have the resources to pursue your goal, you start the interior fire attack, making sure to request additional resources to assist you and stand by for contingencies. If not, you do something else (Figure 13.4). What are your something elses?:

- Wait for help.

- Hit the fire with a master stream until the cavalry arrives.

- Prepare yourself for the potential of a much bigger fire.

Just remember that each of these acts leads you in a different direction. Your original goal was to limit the fire. These actions will not normally accomplish that aim.

It is at this point in the fire that you may be capable of asking and answering your next question: "How am I doing?" To answer this, you must compare the picture of what you wish to see with the fire that sits before you. That is

13.4 If a defensive attack is the most prudent way to contain the fire, order it. Do not risk people unnecessarily. *Courtesy of Harvey Eisner.*

how you make your assessment. It is a simple process to say to yourself, "Is this what I wanted to accomplish?" If the answer is yes, you are doing fine. If it is not, you had better pop the clutch and begin shifting gears.

You may well be saying to yourself right now that what I have said seems too simple to be of any use, but you are wrong. When you are facing the trial of an on-going fireground combat situation, the last thing you need is COMPLEXITY. You need simplicity. You need something that is easy to understand and simple to do with the forces you have at your disposal. Although your attack plan needs to be simple, it must be one that can be grown to fit the needs of bigger fire situations. That is what I am offering you in this book: a simple procedure that has been developed one fire at a time. A major component of a good strategy should be simplicity.

During periods of pitched battle is not the appropriate time to experiment with complex tools. The old phrase, "Put the wet stuff on the red stuff," is a good starting point. And if that is your sole requirement in a simple situation, then the disappearance of the red stuff is a good clue that your goals are being met. All of this leads us to the subject of the next chapter: "Can I terminate this incident?"

14

The Final Step in the Chain: Can I Terminate the Incident?

The last several chapters have been designed to lead to this critical juncture. Unfortunately, too few people pay enough attention to the proper termination of an incident. They merely judge it to be under control and leave. Think of the problems that can occur if you fail to properly assess the scene before terminating command.

Throughout this book, you have been presented a simple, systematic approach to fire fighting. The same way of doing business will be used in discussing how to terminate an incident. Quite simply, there are three things you want to do to the fire you are fighting at any given time:

- Limit fire's spread to the smallest possible area.

- Bring the fire under control.

- Do both of these without injuring any of the firefighters under your control.

REVIEW SIZEUP QUESTIONS

While these are worthy goals, their accomplishment can be a complex undertaking. Remember the series of questions you need to answer in order to size up the fire (Figure 14.). Each of these questions in turn generates a series of sub-queries. Let us take another look at all of them:

1. What have I got?

2. Where is it?

3. Where is it going?

4. What have I got to stop it?

5. What do I do?

6. Where can I get more help?

7. How am I doing?

8. Can I terminate the incident?

14.1 Can your troops bring this one under control safely? *Courtesy of Harvey Eisner.*

In order to reach the final decision, you must ensure that each of the earlier questions has received a complete review (Figure 14.2). Are you sure that you have fully assessed the incident? Have your people reacted to the fire as you thought they should? Was what you thought you had, what you really found as you went along? (Figure 14.3)

Oddly enough, the answer to this question is called experience. Given enough experience, if properly assessed, catalogued, and recalled, your ability to understand the "What have I got?" question can be vastly improved.

The manner in which you and your people were able to work through the "Where is it?" question will tell you if you have found it and stopped it. As a friend of mine once said, you must think of fighting fire as indulging in a game of sorts. As the Incident Commander, you are engaging fire in a battle of wits. Do not enter the contest half prepared.

Both sides in this battle are limited to certain parameters. Mr. Fire:

- Must obey certain physical rules.

- Must operate within a given battleground.

- Can use only such fuel as is provided by the situation at hand to sustain itself.

- Reserves the right to surprise you from time to time (Figure 14.4).

14.3 A large incident may make it difficult to determine what you are really facing. *Courtesy of Harvey Eisner.*

14.2 Have you answered all sizeup questions accurately? *Courtesy of Harvey Eisner.*

On the other hand, you as the Incident Commander must:

- Use an understanding of the physical rules of combustion as best you can to meet, greet, and halt Mr. Fire.

- Seek to gain a better understanding of the battleground.

- Place such resources as you possess in Mr. Fire's way.

- Understand that your chances of surprising Mr. Fire are probably nonexistent.

Please bear in mind that you possess one distinct advantage over Mr. Fire: You have a brain and can think. Unfortunately, we have seen a number of people over the years who operated as though this were an alien concept. In those cases, Mr. Fire usually won the battle (Figure 14.5).

DID YOU ACCOMPLISH YOUR GOALS?

As you approach the termination stage, you must do so based upon a series of positive responses to the following questions:

1. Is the type of incident you thought you had what it turned out to be?

2. Have you found where the fire started, traced its route of travel, and halted its progress?

3. Do you feel that you met the fire and drove it back successfully? Or at the very least held it to the block of origin? Or, failing, that, the city of origin?

4. Are you at a point where you can begin returning people and equipment to service?

5. Are your troops in any condition to return to service?

6. Is it safe to release any mutual aid back to their home towns? A lot of your answer to question five will work to answer this question.

14.4 Remember that even what looks relatively straightforward can surprise you. Don't become complacent. *Courtesy of Mike Wieder.*

14.5 Despite our best efforts, the fire may win. *Courtesy of Harvey Eisner.*

7. Have you done that one last thing to ensure that the fire is extinguished, whatever it might be? (In other words, keep the old brain churning until there are simply no more thoughts to think about the incident in question.)

CRITIQUE YOURSELF

The answers to each of these questions can assist you in two ways. First, they can help you handle the situation you currently face. And second, they can help you hone your future skills as an Incident Commander. The more you work to critique your operations as they wind down, the better will be the quality of your experience. Let us look at the first question. Suppose you thought that you were facing a room and contents fire, and it turned out to be a fire involving several floors. What are some of the clues you may have overlooked?

1. Smoke on all floors is a good clue that you are facing a cellar fire, with the potential to spread to all floors above it.

2. Heavy fire at the rear, which you could not see from the street. All you could see was light smoke, because the bulk of the combustion was in a flaming form (Figure 14.6).

3. Balloon-frame construction allowed the fire to spread up through the walls quickly.

4. An alternative construction method breached the integrity of the platform frame, with the same results (Figure 14.7).

14.6 Heavy fire at the side or rear of a building can cause you to reevaluate your initial answers. *Courtesy of Harvey Eisner.*

14.7 Was there some element of construction that permitted the fire to spread more rapidly than you would have expected?

In tracking a fire back to its source, be sure to start at the outermost point of fire damage. Then have your crews work back to the area of greatest fire damage. You will be retracing the steps taken by the fire. What better way to ensure that you have halted Mr. Fire in his tracks and chased him back to where he started his devastating journey? Another name for this step is overhaul (Figure 14.8.).

The better you get at tracking fire, the more proficient you will become at anticipating where it might travel. This will allow you to operate more quickly and efficiently at future fires. But you must remember what you have seen and done. It's a good idea to keep a set of notes. Over the years you will gain a wealth of knowledge that you might otherwise forget. I urge you to do this because I started one and stopped back in the 1970's. I regret the experiences I may have forgotten due to inattention to details.

By following the path of the fire, you also arrive at the answer to the next question. Did you meet and greet Mr. Fire successfully and drive him back on his heels? Overhaul will assist you in making this determination. So will a meeting with the Fire Chief from your neighboring community, if the fire spread to their city.

An assessment of your rehabilitation efforts will determine the condition of your fire fighting force. Do not rush people back into service (Figure 14.9). Be sure that they are medically cleared if they have been engaged in a pitched battle. Always defer to the decisions of your EMS people. When they release your troops, then you can go about returning companies to service.

14.8 Take time to make sure you have found the source(s) of the fire.

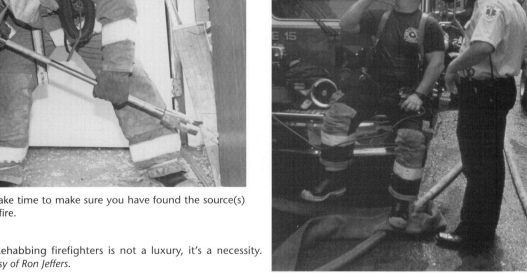

14.9 Rehabbing firefighters is not a luxury, it's a necessity.
Courtesy of Ron Jeffers.

Do not release mutual aid assistance until you have a force sufficient to handle another fire. Tired firefighters can get into a great deal of trouble if they are forced to attend another fire too quickly. I have witnessed this error many times during my career, and it can have serious consequences. When the people are rested, and when the equipment is back in service, release the mutual aid units. And be sure to thank them profusely.

THERE'S ALWAYS ONE MORE THING TO DO

The hardest question to answer is the last. I learned to ask this after viewing a movie that has come to be a favorite over the years. In it, one of the explosives people tells the movie's star that there is always one more thing to do. It seems that there is always one more task to remember, and failing to do it might have deadly consequences. So, keep your brain in gear, and use your eyes and ears to the greatest advantage:

- Listen to what your people have to say.

- Really look for problem areas at the fire scene.

- Look at the appearance and bearing of your troops. (Do they look more tired than they claim to be?)

- Look at what Mr. Fire may be trying to tell you in terms of char, damage, smoke and smells.

My final advise to you is simple, but meant to be thought-provoking. Do not rush away from a fire for something you think is more important. Mr. Fire will take offense and call you back to take up the game again. And in these cases, it will always be on his terms.

15

Fires You Will See Frequently

There are certain types of fire fighting operations that you will see on a fairly frequent basis. While the frequency of each will vary according to the community, these are the ones you are most likely to see regularly. These frequently encountered incidents are the building blocks for more complex operations. For you see, even a complex operation is really many simple operations jammed together at the same time. By using this method consistently and continually, you will become proficient in its application. By becoming proficient in the basics, you will be better able to combat the complex issues. Like any system, the key is to use the same procedures each and every time.

ONE- AND-TWO ROOM FIRES

Every year the trade journals feature a wide variety of spectacular fires. Scores of firefighters labor to operate a wide variety of equipment to handle "The Big One." It makes for great press and it massages our heroic image of ourselves as being all that stands between our world and its destruction. Unfortunately, many fire departments gear their training programs around this scenario. Master streams are developed, and ladder pipes and other elevating platform streams are used to practice for "The Big One." We acquire skills that will serve us well if we do encounter The Big One. In some cases, we even budget for the big one, if we are lucky enough to have the money.

A review of the professional literature tells a quite different story. The large-scale fireground operation is a rare bird indeed. Even in a major metropolitan city, the truly big fire is not a commonplace event. However, what we do encounter regularly are a great many smaller incidents. In fact, most fires require only a single hoseline for extinguishment. I can remember reading many years ago that approximately 90 percent of actual fires were controlled with one line. My three decades of experience tell me the same story.

What, then is the purpose of this discussion, if most of our fires only need a single hoseline? Quite simply, these one-line fires can be challenging and are extremely critical, because if you do not place that single line in the correct place, you will end up with a big fire. And if you do not cover that single line with an appropriate backup line, you are courting disaster. Perhaps the biggest problem we face with these smaller incidents is complacency. We tend to suffer from the "We can handle anything" syndrome. This is a serious matter, particularly in those fire departments with a low fire fighting workload. Complacency can also be compounded by the "one-size hoseline fits all," approach to fire fighting. This is the syndrome where the department only knows how to use one hoseline, regardless of a fire's magnitude.

Many of the firefighters who conducted our in-service training during the early 1970's entered the fire service right after World War II. In those days there were literally just two sizes of hose to use. You either hit the fire with a booster line, using your small tank capacity. Or you laid out with a 2½-inch hoseline, taking the time to make suction at a hydrant, stretch your lines, and go to work. These were the people who then were blessed with the introduction of the 1½-inch attack hoseline. They were so pleased with the increased efficiency, lighter weight, and easier maneuverability of this new hose, that they used it whenever they could. And in some cases, misused it. Many times a fire got out of control because an initial attack line was too small to get the job done. The major lesson we learned from these fine folks was the need for two distinct attributes: flexibility and common sense. In our case, we operated according to an extremely simple rule. "Big fire, big water; little fire, little water."

An astute Incident Commander must not just look at what it going on around him. He must see what is there. While this is true at all fires, it is particularly true during a room-and-contents scenario. The key to success in a room-and-contents fire lies in the sizeup conducted by the first-due company officer (Figure 15.1).

Let us use our simple procedure to assess the problem. The key points are:

1. What have I got?
2. Where is it?
3. Where is it going?
4. What have I got to stop it?
5. What do I do?

In the fire shown on the next page, it is fairly obvious that we are facing a structural fire. So that is the answer to question one. The answer to the second question is also fairly simple: it is coming from windows on the second floor. Answering question three takes a bit more thought. When we ask where it might spread, we are faced with the following thoughts:

- It can move laterally. (To either side)
- It can move vertically. (Upward)
- It can move vertically and laterally. (That is to say, up and across)

Make an effort to ask these questions consciously so that you do not miss an alternative answer. Learn to question yourself as a matter of daily practice. This will allow you to become comfortable with the this style of command.

Your general operating guidelines (GOGs) will assist you with the answer to question four: What have I got to stop it? I recommend that the minimum response to any structural fire fighting operation be made according to the recommendations of the National Fire Protection Association (NFPA). They make reference to the need for two engine companies, one truck company, or a unit capable of doing truck work. They suggest that these units be staffed by a force of 12 fire personnel under the command of a chief officer.

My experience has been that a force of this size will be able to deploy two attack hoselines backed up by a large-diameter supply hoseline. The personnel on the truck company should be able to assist the engine companies in forcible entry, if needed, and vent the structure as needed. They may also be capable of searching a small-area building. After completing these functions, they can control the utilities and assist with salvage and overhaul. This force should be sufficient to control the fairly simple room-and-contents fire shown above.

I strongly recommend that you request a reserve force that is equal to the initial assignment suggested above. You can never call for help too soon, but it is a very common mistake to wait too long to summon assistance. You can always send the extra help back if you do not need it. Remember that there are a number of reasons to call for help:

- Firefighter Safety and Assistance Team units (FAST Team)
- Relief of personnel and rehabilitation
- Unexpected contingencies

In a room-and-contents fire scenario, the basic list of tasks to perform consists of the following:

1. Establish command
2. Size up fire
3. Stretch attack hoselines (Figure 15.2)

15.1 Confining a fire of this type to its area of origin is a challenge you will face frequently. *Courtesy of Harvey Eisner.*

15.2 Always stretch sufficient hoselines to get the job done.

4. Stretch supply hoselines

5. Perform forcible entry

6. Ventilate smoke and gases

7. Perform search and rescue

8. Locate, confine, and extinguish the fire

9. Check for fire spread

10. Perform salvage and overhaul at the fire scene

11. Terminate command

By working according to a checklist, you will be more efficient. The chances of missing something are lessened. And do not forget that a major structural fire is merely a large-scale version of the room-and-contents fire. Yes, it is more complex and dangerous because of a variety of factors. But if you can tell yourself that it is just the next logical extension from the room-and-contents fire, your mind will be better able to work its way through this operational plan.

ATTIC FIRES - DANGER IN THE OVERHEAD

As you roll in to a report of a structural fire, you note smoke off in the distance. As you turn into the block where the address is located, you note smoke and flames coming from a window just under the roof line of the house in question. You are immediately given the answer to the first two questions of our fire fighting system.

1. What have I got?

You have a structural fire in a residential dwelling.

2. Where is it?

It appears to be in the attic (Figure 15.3).

3. Where is it going?

It is going to burn the roof off the building unless you take some positive action to halt its spread. If that were all, you might not need to be as concerned with interior fire fighting. Unfortunately, if left unchecked, it might drop down and then spread laterally. Be on guard for where the fire might be hidden.

15.3 With an attic fire, your objective is to keep it from spreading down into the rest of the structure. *Courtesy of Harvey Eisner.*

4. What have I got to stop it?

Your local operating guidelines should deliver no less than the two engines and a truck company that we suggested above. The same holds true with the need for a reserve force to supplement the attack force.

5. What do I do?

You can fight the fire. Or you can let it burn itself out. I would recommend the first as being less damaging and far more acceptable to the person who owns the home and expects you to protect it for them.

Upon arrival, you also note that there appears to be no fire on the first or second floors. Take the hints that the situation presents. You are facing an attic fire. Be ready for a hot and dirty battle, one that should be approached with a great deal of care. Bear in mind that there are a number of factors that can lead to an attic fire:

- Defective wiring
- Defective chimneys
- Interior fire exposure
- Exterior fire exposure
- Incendiary onset
- Lightning
- Defective attic ventilation fans
- Heating unit problems

Whatever the reason, you may find it necessary to enter the attic to attack the fire at its base. Depending on the needs of the building owner, you may find attic layouts ranging from an unfinished open space to a well-built storage area. It is not out of the question to encounter a full set of living quarters, as well as ductwork and ventilation equipment.

It is also good to remember that the cause of the fire can also lead to its spread. If it is defective wiring, the fire can spread back along the wire run to spread the fire further. A bad chimney can be unsound in more than one spot. And an incendiary fire may have more than one point of origin. Or the material that caused the fire may also allow it to spread by burning through the construction that separates the attic from the rest of the building.

Remember that no two attics are the same, even in a residential development of similar houses. In some cases there will be a single narrow walkway down the center of the attic. In others you may encounter a full floor, with insulation under it, and a carpet above it. In the finished attic, you will also encounter gypsum board walls, and perhaps even some paneling. Fires in finished attics will require a great deal of truck work. Walls and ceilings will need to be opened. In the unfinished attic you will need to be careful as you travel along the walkway. Many an unwary firefighter has fallen through a plaster board or sheet rock ceiling into the floor below.

Regardless of whether the attic is finished or not, you will be forced to approach the fire from below. If you are fortunate, there will be a stairway. Many times, however, you will need a short ladder. This can be placed through the scuttle door to allow access.

Attics are difficult to ventilate. There may be louvers at either end of the attic. Frequently in older homes there is a space between the walls and the roof that allows a fire to breathe. There may also be windows that can be opened or removed, as the situation dictates. These are your best opportunities to relieve the pent-up smoke that accumulates during a fire. Many fire fighting authorities suggest that roof ventilation is not necessary, but my advice is to look at each situation. Use whatever opportunities exist to remove the smoke. Open the roof, if that helps. Open the windows or remove the louvers. The damage you do will be minimal compared to the benefit you will provide.

The best way to extinguish an attic fire is to move in and apply water to the base of the fire (Figure 15.4). You may take a bit of punishment from the heat, but that can be overcome by the power of your attack hose stream. If necessary, have the back-up attack line in place to complete the attack.

A common error I have observed comes when the fire has burned through the roof. It is at this point that someone gets the bright idea to shoot water in, because they can. This drives the heat and smoke downward. It makes entry to the attic impossible, and may spread the fire downward. If the roof burns through, this is good. It allows the smoke and heat to exit the structure through the natural method of ventilation, where heat is allowed to rise and move away. Directing hose streams through the hole in the roof is less than useful. It may in fact be dangerous.

Once the roof is opened, you can move into the attic space and knock down the remaining fire. It is important to remember that you should use as little water as possible. The water that does not vaporize will travel downward and cause water damage. For that reason, you should devote part of the response to salvage efforts. Place the necessary covers in order to limit the damage to personal property in the areas below the fire (Figure 15.5).

15.4 The best way to extinguish an attic fire is to ventilate it, then move in and apply water to the base of the fire. *Courtesy of Mike Wieder.*

15.5 Placing salvage covers will go a long way toward minimizing damage. Be sure your department knows how to fold salvage covers, make water chutes, and so on.

CELLAR FIRES: UP CLOSE AND PERSONAL

By this time, you will probably have arrived at a simple conclusion: Fire fighting is a dirty and dangerous job. We must work very hard at training for those times when the challenge is the greatest. Cellar fires provide just that confrontation. Many buildings are destroyed by fires that start in basements. To understand the problem, you must first visualize the general conditions that exist. Fire fighting in these situations is usually more effective when it is done from an interior stairway. This stairway is often located in the middle of the building. Add to this the fact that there is normally a great deal of smoke and heat rising through that door and you begin to see the severity of the hazard you are about to encounter.

Generally speaking, the contents in a cellar result in a very hot and smoky fire. Usually, the combustion process is slower and there is a lack of oxygen. It is this lack of oxygen that contributes to the density and severity of the smoke, through incomplete combustion. Because building occupants usually do not discover a cellar fire in a timely fashion, these fires tend to be fairly well-advanced by the time the fire department arrives. Compounding the problem is the draft effect caused by smoke moving up through the interior staircase. Firefighters are faced with a situation similar to attacking a fireplace fire down the mouth of the chimney. Many times the discovery of cellar fires is delayed because they begin to burn way back away from where someone might see them.

Because there are not many openings in the average basement, a cellar fire will present a situation where ventilation is limited. If you are fortunate, there might be two doors and a couple of windows. If you are not so lucky, you might find yourself attacking down the ONLY stairway into a windowless basement, or worse yet, a sub-basement. Self-contained breathing apparatus is an absolute necessity. In addition to the standard array of toxic vapors and gases you would face at any normal fire, you have the potential for an oxygen-deficient atmosphere in a below-grade area (Figure 15.6).

Putting water on the fire can be a hit-or-miss proposal. I have often heard it said that you cannot hit what you cannot see. A cellar fire is a prime example of this old axiom. Not only can you not see the fire, but you might not even be able to see the obstructions that can deflect the water away from where it can do the most good. It is often very difficult to determine whether the fire is moving up beyond the reach of your hose streams.

Construction also plays a critical part in your success at attacking and extinguishing a cellar fire. If the building is of balloon-frame construction, your first clue to that effect might be fire on an upper floor. This would hold true for elevator and dumb-

15.6 Cellar fires pose ventilation problems, entry problems, and collapse problems. They are among the most dangerous type of fire you will face. *Courtesy of Metro West (MO) Fire District, Jim Silvernail.*

waiter shafts, and various pipe chases and utility shafts. Anticipate that heat and smoke will do the natural thing and rise upward in the building.

All floors above the cellar are at risk as well as any people in those areas above the fire. You may need to get a hoseline between the fire and the people to allow for a safe exit. You must search and ventilate these upper floors to ensure that everyone gets out. This will eat up a hearty share of manpower resources, so think about extra help early on in the operation.

Cellar fires are often difficult to penetrate with handlines. Therefore, firefighters sometimes must resort to lines from windows and doorways, or cellar pipes from above. However, your best chance at early control and extinguishment comes from an aggressive attack down the interior stairway to the seat of the fire. To do this may require advancing two or three lines abreast into the hottest and smokiest part of the fire, usually the at the top of the stairway leading to the basement. Ideally, there will be a second stairway that can be used to vent heat, steam, and smoke from the basement as your attack advances in on the fire.

Be very careful! If there is no second entrance, the heat and steam from your attack can blow right back into your face. Regardless of how you proceed with your attack, a line must be placed to protect the interior of the building from any fire that seeks to burn up and out of the inside stairway door. If you cannot make entrance into the basement for a quick attack, leave the door to the cellar closed and place a line to protect that point, in case the fire burns through the door before your alternate plan of attack can work. Remember, in some cases, you will be tempted to work above or move over an area involved in fire. **This is not recommended, and should be considered unsafe.** Floors can and do collapse in a short period of time under such circumstances.

What could constitute your alternate plan? In many cases, you will have to resort to opening holes in the floor from a safe area and inserting cellar nozzles or Bresnan distributors down into the cellar space, to control its spread. Such an act may make it possible to enter the cellar. I can remember an incident where this tactic saved a very old Catholic church in Freehold, New Jersey many years ago. Holes were cut, lines were stretched and a Bresnan distributor was placed in the way of the fire's progress. The fire was stopped in its tracks.

As you attack, remember three key points:

- It is critical to ventilate these fires.

- You must provide a second line at the rear.

- Never place your people directly over an uncontrolled fire.

You must let the heat and smoke out. You also need to provide a line at the rear. This line is usually not used to attack the fire because it could lead to opposing streams. But it is available to protect against exterior fire spread.

A cellar fire may well be the toughest job you ever have to face. There is no other way to be ready for one than to learn your business, drill regularly, and recognize the signs that you are facing a cellar fire. Other than flames and smoke coming directly out of the cellar windows, the best clue to a cellar fire is coming upon a building where smoke is coming from every level of the structure. In all my years of fire fighting, I have never seen this clue fail to warn of a cellar fire. When fire and/or heavy smoke are showing on the first floor, it is always a good idea to check below.

Let us look at the our approach as it applies to cellar fires:

1. What have I got?
A structural fire.

2. Where is it?
Below grade.

3. Where is it going?
Up until it hits a blockage and then laterally (Remember that old buildings may have balloon frames that really speed up this trip.) In addition, in certain newer buildings, contractors have been known to violate the integrity of walls in a variety of ways. Be on guard!

4. What have I got to stop it?
What is your standard response and is it enough for the fire you find? Let that be the starting point for your decision-making process.

5. What do I do?
Deploy your forces to best advantage. Call for help. Constantly observe and evaluate the progress of your forces. If there is the least doubt in your mind as to the safety of your people, pull them out and go to a defensive operation. Although we tend to think of ourselves in macho terms, never endanger people for property. We can replace a building, but not a human life.

In summary, cellar fires are tough and nasty affairs. They are an instance where your skill and experience will be called directly into play in a tough and dirty up-close battle. Know these facts and be ready.

VEHICLE FIRES

The incident you will see at least as often as any other involves a fire in a motor vehicle. Generally these are classed as car fires, but you may also be called upon to deal with box van and tractor-trailer scenarios.

The average car fire is not a real challenge. But it this very fact that makes them such a problem. All of us are subject to complacency when it comes to fighting a car fire. Some people skip the self-contained breathing apparatus. There are those who turn out without their turnout gear. And who can forget the booster line buddies who go after a blazing car with the red line? These are just a few examples of common errors committed fighting car fires that can very quickly come back to haunt you (Figure 15.7).

How should you approach a car fire? Carefully, from an upgrade position, preferably upwind. In case there is a release of flaming product, you want to watch it run away from you, rather than toward you. Approach from the corners of the vehicle to avoid injury from exploding bumper shocks. I believe in attacking vehicle fires with a minimum of two 1¾-inch handlines. One line serves as the primary attack line, while the other is a backup, just in case something goes wrong.

Let us take a look at our sizeup procedure. Base your assessment on the picture above.

1. What have I got?

A fire in the trailer section of a tractor-trailer unit (Figure 15.8). But what have you really got? Placards or markings may give you a clue. Until you can review the shipping papers, however, you just will not know.

2. Where is it?

Somewhere in the midst of a jumble of cargo in the center area of a commercial trailer unit. Perhaps the driver can assist you by telling you which part of the load contains what materials — that is, if he knows. Many times the driver is not there when the cargo is being placed on board. However, it is better to talk to the driver than not.

3. Where is it going?

In the absence of an exposure problem, you might not have anything more than a mobile room-and-contents fire. But you must be able to identify the contents of the room, if you are to avoid blowing yourself to Kingdom Come.

4. What have I got to stop it?

Whatever your standard response is to a large vehicle fire. In some fire departments, a vehicle fire receives a single engine company response. We have found this to be inadequate. At a minimum, I suggest that your response should be an engine company, a truck or service company, and a chief officer for large vehicle fires such as trucks, vans, mobile homes, and the like. In areas where there is a water supply problem, a second pumper or a tanker pumper should be added to the response.

5. What do I do?

Deploy your forces to best advantage. Place your units upwind and upgrade. Stretch two handlines of a minimum 1 Clv-inch size. Ensure that all personnel are wearing full personal protective equipment and are equipped with SCBA. If hazardous materials are obviously involved, do not commit

15.7 Although car fires look simple, be sure to wear proper protective clothing and take no chances; the primary dangers are to occupants and firefighters. *Courtesy of Mike Wieder.*

15.8 Always be alert to the possibility of hazardous materials or flammable liquids when you deal with tractor-trailers. *Courtesy of Harvey Eisner.*

your people to positions of exposure from solid, liquid, or gaseous materials. When in doubt, isolate the area and treat it as a hazardous materials incident.

Remember, that if you think you need help, call for it. Do not attempt to be a hero with other people's lives. In those cases where you find that the reported vehicle fire is in a structure, call for a full structural response. I have been on incidents like this a number of times. It is a real surprise, but a calm reaction from you can keep things from turning into a disaster.

It is important to train your people in the latest technical and safety improvements. Air bags, hydraulic bumpers, and power equipment present challenges that we of an earlier generation did not face. It is not the purpose of this book to teach you how to use such tools, but rather to call your attention to the fact that there are many points to ponder. You can find out a great deal by talking to an auto repair person, an auto body repair specialist, or a technical rescue technician. This is a case where a few moments spent in learning can prevent a lot of time recovering in a hospital. The simple act of wearing SCBA may well prevent a great deal of future suffering.

Vehicle fires are common. Just don't allow yourself to become complacent about them. Drills spent learning to control these types of incidents can pay dividends.

ELECTRICAL FIRES

In any given year, you will respond to a number of incidents that end up being classified as electrical in nature. These range from shorts in kitchen appliances and in ballasts of fluorescent lights, to extension cords run under carpets, right on up to full-fledged blazes in electrical power stations. These fires all have one thing in common: the dangers inherent in electricity. In addition, an overheated ballast can burn the hand of any unwary firefighter who chooses to grab hold of it without wearing the necessary gloves. The role of a firefighter is quite simple in these situations. You need to make the unsafe situation safe. Your questioning sequence should go something like this:

1. What have I got?
That unmistakable smell of a burning ballast.

2. Where is it?
Here is the hard part. You must inspect each of the ballast units to find the one that is involved. This is not a problem in someone's home. But if you have been called to the local shopping center or to a large supermarket, then you will have a real problem finding just the right fixture.

3. Where is it going?
If you do not find the involved fixture, it can short out back through the building electrical system. Then you will have a big fire and need big water.

4. What have I got to stop it?
You may have less than a full assignment, because in many cases a single engine will be sent out to investigate. If you need more help, call for it. Do not let your pride lead to the destruction of a structure, because you do not have enough people to conduct a proper investigation.

5. What do I do?
There are just a couple of simple tasks to keep you out of trouble:

a. Find the problem fixture

b. Remove the tube or bulb

c. Cut off the power to the fixture

d. If it is an older model with the separate ballast, remove the ballast and dispose of it in a safe place. Do not simply toss it into the trash, for it may retain enough heat to cause a fire.

e. Advise the property owner or tenant to have the fixture inspected by a licensed electrician.

While this whole section might seem to be simple, it is the sort of situation that can get out of control. It is a case where your knowledge of the basics can allow you to provide the best possible service to your consumer, namely, the citizens in your community.

BRUSH AND TRASH FIRES

Whole texts are devoted to fires involving the wildland/urban interface. I am not thinking here of the large-scale brush and wildfire scenario. That is a specialized field. I am interested in the more modest situation involving a small stand of brush adjacent to a residential occupancy. This is the type of situation faced by both urban and suburban firefighters. The objectives are quite simple:

15.9 Sometimes the traveling dumpsters catch fire, too. Make sure the proper authorities are alerted for traffic control and cleanup. *Courtesy of Mike Wieder.*

- Do not injure a firefighter

- Limit the size of the fire

- Prevent extension to a structure

The main task is to place water between the fire and the area of exposure. Many fire departments use a booster line for this type of operation. They say that the booster line is easier to move and it supplies enough water. My experience tells me otherwise. I believe that you should operate at these sorts of fires with a 1¾-inch hoseline, equipped with a variable stream nozzle. The hose is more manageable and the flow will get the job done. As a matter of fact, I have had tremendous success with a prepiped deckgun to stop a fire in its tracks. While there are those who say that this is overkill, it is simple, direct, easy and effective. Give it a thought.

There will be times when you may have to exert a bit more care and effort to get the job done. If there is a bit of wind and the distance between fire and exposure is small, you will face a greater challenge. Always remember that the primary concern in these cases is **FIREFIGHTER SAFETY**. We would not like to tell a widow that her husband died saving a pile of garbage bags. It just does not make sense to endanger people for property.

Many times I have supervised operations at industrial dumpsters (Figure 15.9). After determining that hazardous materials were not involved, I usually ordered an open-butt supply hoseline to be stretched. The dumpster was then flooded. The object, of course, was to extinguish the fire in the safest manner possible. It is not the fire department's fault that there is a fire. Our job is to extinguish it with a minimum of effort and danger. The removal of a damp dumpster is the vendor's concern.

Let us close with a couple of simple rules. When combating small brush and trash fires:

1. Keep it safe

2. Keep it simple

HOW TO HANDLE THE SMELLS AND BELLS CALLS

One of the biggest challenges facing fire departments all around the world is the simple Smells and Bells stuff. Rare is the case when you will encounter a fire from an automatic alarm response. But you must be ready, just in case. This type of run can be the classic, institutional case of the Boy Who Cried Wolf. You get called to the same place time after time for alarms. After the third or fourth call you start to slow down a bit, you pay less attention and lose your focus. You begin to let your guard down. And in some cases, your personnel response levels suffer. But be careful: I can remember an alarm call that escalated into a raging warehouse fire. After 10 hours of hard work, the fire was finally declared out. As we stood among the rubble, it was tough to remember that the initial call was for an alarm sounding.

The key to success in these situations is to establish a routine that is followed by all responding personnel. How you tailor your response level is a matter of local choice. Many departments tailor their response to the occupancy. Some small departments send a complete response for just about every call that comes in. In larger cities, a chief, an engine, and a truck company may be dispatched to an automatic alarm.

My view is that the size of the department and the size of the occupancy have to be considered in the response equation. The danger to a school is much greater during the period when children are present, so a greater response is justified. An individual development or tract home could receive a smaller response. You must decide.

Carbon monoxide detector alarm runs are on the upswing. Just as the installed residential smoke detector has generated an increase in runs, so has the installation of residential carbon monoxide alarms. Many fire departments blow these runs off as nuisance calls. Others are not equipped to handle them.

My department has a sizable number of such incidents, and I believe that carbon monoxide detectors serve a purpose. However, the fire department must respond in a studied manner using a standard operating procedure. The object is to operate using carbon monoxide and oxygen level detection equipment.

By establishing a procedural checklist, you ensure that the basics of a thorough search are listed. The firefighters can then move through the building. By using the check sheet to guide their travels and the detector to meter the air, they can make a thorough search of the structure. There are some rules to follow:

- All personnel will use full personal protective equipment (PPE) and self-contained breathing apparatus (SCBA).

- All personnel will work in two-person teams.

- A two-person safety team will be maintained outside with full (PPE).

- Personnel will use the appropriate checklist to guide their search.

- Personnel will use CO detectors to check for the presence of carbon monoxide.

- Check lists will also contain questions to see whether people have been exposed to CO.

For additional guidance on the development of a CO detector response program, you should contact a manufacturer of these devices. They can offer excellent advice, which will help you do your job: saving lives. It's up to you to develop the best possible plan, drill on it, and then use it continually.

SUMMARY

The incidents you respond to most frequently are not usually complicated ones. Nonetheless, it is important to use the same system each time. By approaching each incident with the same array of questions, you can approach each event and work through it in a systematic way. As you gain skill in using this system, you will be able to generate additional questions of your own that help in the decisionmaking process. The answers to these questions will guide you in performing your job safely.

16

Fires You Will See Occasionally

This chapter moves from the fires you are most likely to encounter to fires you are likely to see only occasionally. I have based the list on my experience in both the urban and suburban environs. The fact that they come less frequently means that you need to concentrate on preparing for them. You will also need to review them on a continuing basis, to ensure consistent familiarity.

TOWNHOUSES

Whenever you speak of townhouse construction, you should also lump the garden apartment and the condominium complex into the same paragraph. Each of these is designed to create housing for a large number of people in a limited area. These are the modern outgrowth of the old row house, which is found in cities all around the world (Figure 16.1).

Problems:

- Congested parking limits fire apparatus and hydrant access

- Mixed construction

- Many people in a small area

- Severe internal and external exposure problems

- Transient residents who may not take care of their individual units within the complex

- Limited access to the individual units within a complex

- Heavy nighttime occupancy load, coupled with a variable daytime population

- Rarely feature automatic suppression systems

- Modern lightweight wooden truss roofs may eliminate top-side ventilation possibilities

- Delays in finding and accessing particular addresses

16.1 Townhouses often have access problems; they always have exposure problems.

In your use of our sizeup methodology, you may be able to base many of your decisions on the same model used at a single-family residence (see room-and-contents fires). However, a primary difference comes in the area of exposures. You must be concerned, early on, with where the fire might be headed. Fire in these buildings can extend internally and externally. And it can happen very quickly. Anticipation can be a lifesaver, both literally and figuratively.

The weather aspect must also be addressed early on in your operation. The difference between wind and calm is critical in these situations. Further, you will be faced with exposure problems, which can range from the units on either side of the fire area, to units above, below, and nearby. If high winds are present, entire complexes may be placed at risk. In situations like this you must quickly follow the "What have I got?" question with "Where is it going?" By staying ahead of the fire, you will have a better chance of success in controlling fires in townhouse complexes.

The cul-de-sac style street layout preferred by many developers will make it difficult for you to deploy your resources. You may have to lay a large-diameter supply hose into a cul-de-sac from the nearest hydrant and utilize multiple attack lines from that unit. Because your ability to supply water in sufficient volume is very important, allow for multiple means of approach in your pre-incident planning. Because of the potential for fire spread, mutual aid should be built into your response. **Beware: Fires of this type can quickly get out of control if you underestimate their potential and fail to call for help in a timely fashion.**

STRIP MALLS

When we refer to the strip mall, we speak of a collection of commercial stores under a common roof. They are usually located along busy highways. These malls are a product of the post World War II suburban boom. They are the suburban equivalent of the commercial taxpayer building found in the larger cities. Strip malls are usually built quickly and inexpensively, and are built to maximize the use of available space (Figure 16.2).

Problems:

- Access may be congested.

- Parking may restrict approach to individual units.

- Water supply may be located at a distance.

- Access to hydrants may be limited by parked vehicles.

- In many cases, noncombustible buildings are tied together with decorative wood facing, overhangs, and connectors.

- Occupancies can change frequently, in some cases literally overnight.

- Mix of occupancies leads to a mix of hazard levels.

- Variable life hazards.

- Roof construction (lightweight pan decking with steel bar joists, bowstring truss or lightweight wooden trusses) may eliminate ventilation opportunities.

Your primary concern in these occupancies is life safety. As always, the safety of your firefighters should be foremost in your thought processes. Because there are those times when you would not expect to find people in the building, you must carefully weigh any commitment of personnel. **Never risk people for property.** On the other hand, when you would expect people to be involved, you have to carefully review the fire conditions to see if it is safe enough to place your people in harm's way. Remember, your primary concern is for your fire personnel (Figure 16.3).

Such factors as volume of fire, condition of structure, color of smoke, and the reliability of occupancy information should form the basis for your decision. Once you have decided to commit people, monitor conditions continually. Maintain solid communications and require periodic reports. And if your gut tells you that commitment of personnel is wrong, go with that feeling. Better to err on the side of safety. With this type of fire your "What do I have?" question can be quickly overridden by the "Where is it going?" scenario. With the potential for spread through hidden spaces you must remain a step ahead of the fire, or you will surely lose. I have seen fire jumping two stores ahead. In one case, the fire spread through a pipe chase. In another, it jumped ahead through a boxed-in wooden marquee. These were cases where the fire had a good hold on the building before we arrived. We were successful in halting further spread because the truck company members were performing a thorough reconnaissance of the buildings. They were doing this as a part of their primary search. Since these fires occurred during times when people might have been in the buildings, this was acceptable. Let me repeat one more time: People are your most important resource. Do not waste them. Do not damage them unnecessarily.

16.2 Strip malls house a variety of occupancies and hence, a variety of hazards, under one roof.

16.3 If you are not faced with a rescue situation in a strip mall, make sure that your firefighters are not placed in harm's way.

Because it is advisable to limit the rooftop exposure of your people, you may need to use alternative means to remove smoke from a building. Positive-pressure ventilation, when properly used, is appropriate for these circumstances. If access is available from two sides, cross-ventilation is appropriate. However, large, complex building layouts might require a combination of these two styles of smoke removal. These buildings are not likely to have air handling systems that are sophisticated enough to remove smoke and heat.

It might seem like a tough thing to say, but there are times when the best thing for everyone concerned is to hope that the fire burns through the roof. Any firefighter whom you would risk sending to the roof must be working directly from an aerial device preferably attached by a safety belt to a ladder or bucket position.

In strip mall scenarios you may have difficulty answering your "How am I doing?" question. This difficulty arises from the multiple occupancies in a common area. You may have to do a bit of damage to adjoining occupancies in order to identify how far the fire has spread. In this way, you can work back into the damaged area from an undamaged area. This is a critical part of the overhaul function. After determining the extent of fire spread, you must then ensure that nothing remains smoldering, which can rekindle the fire you worked so hard to contain. You should also ensure that any evidence of illegal activity is preserved, should it be encountered.

Alterations can make a building weaker, so watch for signs that new construction has been mingled with old. There have been cases of structural collapse where an existing building was weakened through the addition of steel I-beams to replace heavy, older style timber construction.

NEIGHBORHOOD COMMERCIAL OCCUPANCIES

This is a wide-open category, and there is no one basic type of neighborhood commercial building. You will encounter as many different kinds of commercial occupancies as there are types of businesses (Figure 16.4 a - c). What works in a given community is what is imitated.

What you will encounter is the product of the minds of enterprising local business people. They discover a need and then develop a structure to meet that need. For that reason, you must become familiar with the general characteristics of this style of building. In this way, you will be able to apply the appropriate response to the problems you will be called to handle.

Problems:

- Life hazard of occupants above the commercial area. (Potential 24-hour-per-day exposure)

- Life hazard of employees and customers. (Exposure varies by time of day)

- Wide range of storage-related difficulties. (May vary by the season)

- Wide range of materials that may be in storage at any given time.

- Depending on the age and construction of the building, there may be no fire walls to limit fire spread.

- May be very difficult to ventilate.

- May not be large enough to require installed automatic fire suppression.

16.4a City-style strip malls often have the additional problem of people living in apartments above the stores.

16.4b Smaller business occupancies like this dry cleaners can be found just about anywhere.

16.4c Family-style restaurants also proliferate.

- Parking may limit access.

- You may be so familiar with the occupancy in question that you overlook the hazards. (Particularly true with fast-food establishments.)

You must become familiar with the many different styles of neighborhood commercial buildings. Whether it is during the course of regular visits as a customer, or in your role as a firefighter, pay attention. If you spot something out of the ordinary, make note of it and pass it along to your fellow firefighters. If it seems to be a violation of the appropriate fire codes, be sure that the proper authorities are notified. A fire prevented is one less that can reach out and bite you.

The key to success in many neighborhood commercial fire situations is ventilation. However, I urge you to be extremely cautious when considering rooftop ventilation. Many of these newer struc-

16.5 In commercial structures, positive-pressure ventilation is often the safer way to clear out smoke. *Courtesy of Mike Wieder.*

tures are built over lightweight wooden trusses. Or they are lightweight metal pan decking resting upon metal supports. And even if they are properly built into the bearing walls of the building, the newer materials can fail quickly.

More thought must be given to the use of positive-pressure ventilation as the safer alternative (Figure 16.5). But to make this work well, there must be a way in and a way out for the air flow. If you push the smoke into the building and it bounces off the rear wall back into your face, you will have a tough time entering the fire building.

There is no set size for a fire in a neighborhood commercial occupancy. Remember this primary rule:

Big Fire = Big Water; Little Fire = Little Water

If you decide that one line is sufficient, be sure to back it up with another of equal or larger size. If you can blast the fire with a deck gun or a 2-½-inch solid bore handline, do it. Do not be a hero.

Always anticipate where the fire might be going. If it is down, it can go up. If it is in the building on the left, it can move to the right. And ponder just how many ways you can work this equation. Fire can even drop down through the inside of certain types of walls. Be ready. Most important, be prepared to meet the threat with sufficient force. To do this, you need knowledge. To gain knowledge, perform thorough pre-incident planning. You should do this to protect your people. Remember the old saying: "Never go to a gun fight with a knife."

You may be tempted to act to help a friend who owns or operates the store. Be patient and operate deliberately. Do not throw a lot of water just to impress the public. Do not race into a building just to please the owner. Conduct your size-up. Plan your work and then work your plan. Do not be afraid to open up the building, or for that matter, surrounding buildings. The fire may be moving faster than you think. If you do not anticipate, you may experience a larger than necessary loss. The primary goal, as always, is safety.

SMALL HOTEL/MOTEL OPERATIONS

In cities, towns, and villages all across America, you can find a wide variety of small hotels and motels. They may be a mom-and-pop motor park operation like the hundreds that sprang up after World War II. These usually feature a series of cabins around a central office (Figure 16.6 a and b).

Another style is the small-town strip motel. This consists of a single building with the office and a number of rooms all under one roof. These can range from the very simple to the moderately luxurious.

In many places you can find the small residential hotel, which is a holdover from an earlier era. Unfortunately, these often house a wide range of social transients who may be in poor mental or physical health. This creates a serious challenge for fire departments because the residents may be difficult to locate and evacuate.

And do not forget the chain motels. Many are being built in small towns that just coincidentally adjoin the interstate highway system. While their owners may have taken pains to keep them in decent shape, many of them are more than 30 years old. Remember that when pre-incident planning them. They might be built to a far different code than that with which you are familiar.

The life hazard in smaller hotels or motels can be quite flexible. Many times it depends on the traffic flow in the area, the season of the year, and the relative popularity of your community. Pay attention to the traffic in the parking lot. If you roll in on a reported fire at 3:00 a.m. and find a full parking lot, think a couple of thoughts:

- Lots of search and rescue opportunities

- Call for help

Many large, old residential properties have been converted to bed and breakfast inns. These establishments present a growing challenge to fire departments everywhere (Figure 16.7). In these structures you will face an increased life hazard due to the guests' unfamiliarity with the means of egress and the fact that these structures were built for single-family use. In addition, certain inn operators have a limited knowledge of safety codes and rules. If you have any bed-and-breakfasts in your community, get to know all about them. If you are called to one, you will have only a short time to do a number of things:

- Rescue people
- Confine fire
- Extinguish fire

And you had best do them all correctly.

Problems:
- **LIFE HAZARD CAN BE EXTREME.**
- Conditions may exist for panic among occupants.
- Tenants may be unfamiliar with surroundings.
- Hoseline stretches may be long.
- Parking may block access. (both legal and illegal)
- Hydrants may be blocked.
- Water supply may be limited in smaller communities.
- Fire may block unprotected vertical openings, namely open stairways.
- Fire on lower level can expose a great many people.

There have been a number of fires in smaller regional hotel/motels that have resulted in injuries and deaths. The problems listed above make the standard job of fire

16.6a One style of smaller motel features a series of rooms near a central office.

16.6b Even smaller hotels house office and guest rooms under roof.

16.7 Bed-and-breakfast hotels have become very popular; in an emergency, the life hazard can be extreme because guests are unfamiliar with the means of egress.

control just that much more difficult. As always, your primary concern at fires in these hotels and motels is **LIFE SAFETY**. In many cases, the value of the building means very little when compared to the potential for human tragedy. Let me state once again that the more you know about a building like this, the better will be the quality of your decisions when something goes wrong.

Hotels and motels carry high occupant loads. While you are attempting to make entry, startled occupants are attempting to flee through the same areas that you are using. Be prepared for the rush. You must also recognize the need for a solid interior fire attack, wherever possible. A single well-placed hoseline can save a lot of people. I can remember a multiple dwelling fire where a single engine company, using one 1-¾-inch hoseline, shepherded more than 20 people through a fairly heavy fire condition. All arrived safely in the yard outside. (Note: Please note that this single hoseline was one of many in operation. Never operate with less than two attack lines backed up by an adequate water supply.) Remember that your greatest chance for success comes from a team approach to fire fighting. Some people have to do water things, other people need to do break and open up tasks, while still others need to get the occupants to safety. In particular with hotel/motel scenarios, you ignore any one of these at your own peril.

If you can be sure of the vintage and construction of the building in question, it may be possible to perform top-side ventilation. You can make an evacuation stairway a lot safer if you can open up the roof above it. That is one of the primary reasons for top-side ventilation. You need to relieve a building of pent-up smoke, thereby making it safer for the building occupants (Figure 16.8). You can open any roof doors or scuttles. If need be, you can open up the roof with power saws or axes. Be sure to take a pike pole or hook along to poke down into the hole and remove all obstructions.

Here is a case for the use of a strong interior attack to save lives. You may need to operate more than one line at a single position to gain control of the fire. And you may need to use multiple lines, one covering for the other, as you advance down a burning hallway.

In fires of this type, the need for solid command and control procedures is paramount. The Incident Commander needs to know where the troops are. Further, simultaneous attack crews operating on different floors require constant monitoring. There is so much going on that an unwary IC can lose people in the shuffle. By concentrating on the basics, you stand a much better chance of successfully controlling this type of fire. Think about search and rescue, people control, hose work, and ventilation. Getting the people out of dan-

16.8 If it can be done safely, ventilating from the top will help trapped occupants and their rescuers. *Courtesy of Mike Wieder.*

ger is your primary task. Keeping your people safe while doing this is an equal concern. A solid Personnel Accountability System (PAS) can help. By monitoring crews as they are assigned, contact can be maintained by sector commanders, as well as the IC (Figure 16.9).

If you know where the troops were detailed and continually monitor their operations, you can stay on top of events. If something goes wrong, you can start looking where the people are supposed to be located. It beats the heck out of guessing. If all of these steps are being taken, coordination will be improved. And it is effective command and control of the operational troops that spells the difference between success and failure. In these situations, your sizeup may well look something like this:

1. What have I got?

A fire endangering people who may not know much about where they are located.

2. Where is it?

A fire-specific variable that must be quickly answered. Until you know exactly where the fire is located, and its extent, you cannot make proper decisions as to just who is in danger. This lack of knowledge also stymies your ability to take proper actions.

16.9 Have a system for knowing where your people are to improve coordination and increase safety. *Courtesy of Ron Jeffers.*

3. Where is it going?

The type and age of the fire building will be your primary clues here. Older buildings, with minimal fire resistance, will allow the fire to spread. Newer, code-compliant structures may help you by limiting fire's spread. Remember that fire can move up or down as well as to the left or right. As always, anticipation is the key. Based on the circumstances at hand, as modified by your experience and training, decide where the fire is headed. And then meet it there.

While all of this is going on, remember that people are in danger. Rescue and attack can be accomplished at the same time if you have enough people. Do it.

4. What have I got to stop it?

Your response assignment to smaller hotels and motels should be at least as follows:

- Four pumpers

- Two aerials, if available

- Rescue services vehicle

- Sufficient tankers to meet the necessary fire flow, if they are your means of water supply

- EMS units

- 20 to 30 firefighters

If you cannot generate this response from your own forces, then build automatic aid into your program. These are not fires where you can sit back and call your help in a piecemeal fashion. You must save as many lives as time and circumstances permit, hit the fire hard, and limit its spread.

5. What do I do?
You must do the following:

- Rescue

- Exposure protection

- Locate fire

- Confine fire

- Extinguish fire

- Force entry as needed

- Vent as required

- Overhaul as needed

- Salvage as the opportunities arise

6. How am I doing?
If the occupants are saved and the fire is extinguished, you have done extremely well. However, never forget that the difference between this result and anything less is a combination of hard work, training, experience, and fate. Be sure to control the three that are within your power.

SUMMARY
Basically, the types of fire incidents that firefighters are likely to encounter on an occasional basis, are a larger version of the room-and-contents fires discussed in the preceding chapter. The skills you will need are the same; however, you will need more people and you will encounter a greater potential for peril. In this chapter we looked at the problems involved in operations in townhouses, strip malls, neighborhood commercial enterprises, and smaller hotel and bed and breakfast facilities. The common thread in these types of incidents is people. The variables are construction, storage, property maintenance, and weather conditions. Each can be a help or a hindrance. It is up to you to learn as much as possible about each during pre-incident planning.

17

The Big Ones: Fires You Will See Only Rarely

In this chapter I am going to ask you to think big, about incidents that most communities experience only rarely. In some major cities, these incidents occur with greater frequency. Because most of you do not live in large cities, however, you may get very little experience in dealing with large incidents. This chapter is designed to equip you with knowledge for "The Big Ones," – fires in shopping malls, high-rises, churches, lumberyards, schools, warehouses, and industrial facilities (Figure 17.1).

17.1 It is not often that most departments have to cope with a fire of this magnitude. *Courtesy of Harvey Eisner.*

SHOPPING MALLS

These major mercantile occupancies are a relatively new phenomenon. As a lad growing up in New Jersey, I shopped in downtown areas. In 1961, the first mall opened in our area. It has grown by leaps and bounds over the years, and is now like a small city under a single roof. The trend in malls has been for bigger, bigger, better, better (Figure 17.2). There are malls that have become tourist stops because of their size, variety of outlets, and the sheer novelty of the entertainment and amusements these co-located areas.

Problems:

- The primary problem in these occupancies is people. There are likely to be large num-

17.2 It seems as though the large shopping mall can be found almost anywhere nowdays.

bers of people who are unfamiliar with the exit facilities of the mall they are visiting. There are malls that I visit frequently, but it is only because of my background that I look for all of the exits.

17.3 Although pedestrian traffic may be provided in many malls, access for emergency vehicles is likely to be difficult.

- Access can be a problem. Some malls are surrounded with literally acres of parking facilities (Figure 17.3). And not everyone is capable of reading (or willing to heed) the signs that state, "No Parking — Fire Zone."

- Because of their complexity, shopping malls can be manpower traps of the first rank. In order to operate at these facilities, you must visit them frequently and know where the built-in fire protection assistance is located. And without a functioning personal accountability system, you are gambling that you will know where your people will be operating.

- These facilities generate an inordinately large number of "smells and bells" calls. People tend to become complacent. And complacent people get into trouble.

- Certain seasons of the year, especially Christmas, can generate tremendous crowd problems. Many times holiday throngs can border on the occupancy load limit. You need to be aware of what is happening in your mall if you have one in your response area. Periodic visits are a must.

- Certain weather conditions can generate extra crowding. A rainy summer day will cause people to go shopping rather than pursue a recreational endeavor. The same thing can happen on snowy days. Odd, indeed, but true.

- Certain times of the day and week can generate higher levels of pedestrian traffic. Examples of this might be in the evening and over the weekend. Shopping has become a great hobby for a many people.

- The large area of the mall structure will force you to combat the fire using building systems. You will need to know how to move water in and remove smoke from these large, open commercial occupancies. Elevated streams from aerial devices may be incapable of reaching the center areas of a large building or series of connected buildings.

- In malls that have grown over time, you may be faced with differing types of construction (Figure 17.4). There may also be voids, dead spaces, and open areas where fire and smoke can move unknown to the fire department. These potential problem areas must be identified during your pre-incident planning sessions.

Conditions may worsen if older wooden buildings are incorporated into newer construction and the wood is hidden. Trust me when I tell you that you will be among the first to find out if this is happening. However, since it will be hidden, you may not at first guess what is occurring. If you encounter heavy pockets of fire, think wood. And always think combustible storage because nearly all of what is sold will burn at some level of intensity.

One element that you should have in your favor is that most building codes require installed fire suppression in large commercial shopping malls. If a new property is being built in your area, be sure that code provisions for automatic sprinklers and alarm devices are enforced. Once the mall is built, work with your fire prevention personnel to ensure that such devices are properly tested and maintained. Automatic sprinklers can be lifesavers in structures of this magnitude. You and your personnel must train continually on how to support the installed protection. What a shame if you failed to feed the system siamese and the fire beat the sprinklers. Shame on you.

17.4 Malls that have "grown" over time often contain a challenging mix of construction types and styles.

If you do face a fire in a shopping mall, panic among shoppers will be a primary concern. People unfamiliar with their surroundings may race, blindly, for where they think they saw an exit. You will be entering where they are leaving, and will probably feel like a salmon swimming upstream to spawn. Under these circumstances, it is critical that you act in a calm and rational manner. If people sense that you are panicking, it will heighten their sense of fear. The loud, firm voice is much better than a high-pitched whine when barking instructions to evacuees.

These are the types of fires where teamwork, supplemented by strong command and control procedures, is essential. The buildings are complex and the number of areas to be searched are numerous. As the Incident Commander (IC) you will need to establish early control over a number of things:

- Where your people are assigned
- How long they have been on air
- When are they due to be relieved
- The number of possible occupants (very difficult to determine)
- Injuries and casualties as they occur
- Maintenance of a strategic reserve of people and equipment

The use of an incident management system is imperative (Figure 17.5). The quality of the information and guidance from your sector commanders will serve as the basis for your success or failure as the Incident Commander. Given the complex nature of large-scale mall incidents, you must delegate a wide array of command and control tasks on the fireground. Delegate, but do not lose control.

What are some of the other areas of concern in a shopping mall operation?

- Stockroom areas
- Restaurant areas

- Rubbish removal areas and systems

- Rooftop structures

- Potential for fire in heating, ventilation, and air conditioning ducts

As the operation proceeds, each of these must be considered. Actions may need to be taken to ensure that fire has not spread to these areas. Remember that it will take a great deal of time to attend to the many details at an incident of this magnitude.

The need for self-contained breathing apparatus will be great, so early on in the operation you must arrange for a strong air supply operation (Figure 17.6). The need for medical monitoring and rehab facilities goes hand-in-glove with the air resupply situation. The scale of the fire and the feeling that much needs to be done can cause your people to inadvertently overexert themselves. They will respond to the citizens by wanting to deliver a superhuman effort. Your job is to keep this in check. Guidelines exist for monitoring, resting, and re-hydrating your personnel. Be sure that these policies are in place and enforce them.

Should you encounter a fire in a mall, expect a protracted campaign. You will need to arrange for additional personnel to ensure that firefighters are rotated in and out of the fire building on a regular basis. Your logistics will surely involve food, water, fuel and reserves of people and equipment. Plan for this.

17.5 There are many areas to keep track of during a shopping mall incident. An incident management system can literally be a lifesaver.

17.6 A major fire will require more resources, such as a great deal of air resupply, if the troops are to do their jobs. *Courtesy of Ron Jeffers.*

HIGH-RISE BUILDINGS

Problems:

High-rise fire control problems tend to fall into four distinct areas:

1. **Construction**
   a. Steel columns and girders
   b. Sprayed-on insulation
   c. Lightweight floor trusses
   d. Central core tied to columns in outside walls
   e. Central core has all utilities, exits elevators and machinery located in it

2. **Occupancy**
   a. Residential - Condominium
   b. Residential - Hotel
   c. Commercial
   d. Offices

3. **Design**
   a. Must meet codes
   b. Codes are a minimum requirement

4. **Workmanship**
   a. Nothing is ever 100 percent correct.
   b. The people who build high-rise buildings are human: they have good days and bad days.
   c. People can make a lot of mistakes on their bad days.
   d. We will not normally find these mistakes until there is a fire.

Access

We will group the first round of problems associated with fires in high-rises into a topical heading generally called access. These are the nuts and bolts issues that deal with how we get to where our help is needed.

Let us first look at the basic fact that the height of these buildings is almost always beyond the reach of even our tallest ladders (Figure 17.7). This one fact alone eliminates many of the basic tactical evolutions used by truck company members to assist in combating a fire. Because you cannot raise a ladder and climb through a window, you are basically limited to a minimal number of building entrances. This same fact also applies to the number of ways that you can enter a floor to begin fighting a fire (Figure 17.8). You are limited in the ways in which you can approach and attack a fire of any magnitude.

Moving in on a high-rise fire is a very hot, dirty, and dangerous business. Combined with the problem of area is the matter of numerous partitions and obstructions slowing your attack. Rather than building walls to separate work spaces, a great many firms break the floor area into functional spaces by using specially designed partitions. Many of these dividers have bookshelves, electronic office equipment, and large volumes of combustible materials. The fire loading in many high-rise office buildings is quite excessive.

One last fact that you must keep in mind is that there may well be no mechanism to easily remove the heat and smoke from the fire floor. The heat will come at you as though you were entering the door of a blast furnace. If the Incident Commander is unable to arrange for evacuation of the smoke and heat, or if no such mechanism is built into the building, you may be forced to resort to the use of portable, high-volume water application devices to project water into the burning office space.

Size of Structure

The next round of problems will come from the size of the area that you will be forced to face as you move in to combat the fire. Generally speaking, modern high-rise buildings contain a number of large, open areas. You will also be confronted with a wide range of exposed combustible materials (Figure 17.9). These two conditions may combine to confront you with an uninterrupted spread of heat and smoke. You may even experience the phenomenon of fire circling around and attacking you from the rear.

Once you and your people are able to reach the level of the fire incident, you may be faced with another serious problem: The dimensions of the floor may be so large that fire fighting streams have a hard time hitting the actual fire. This will make it necessary for you to either move in and attack with hand-held hoselines or resort to the use of heavy streams supplied by multiple standpipe outlets.

Because high-rises are so large, you will need to mount large-scale operations to allow for an effective fire fighting operation. As a matter of fact, you may not be able to attack until you have staged sufficient resources at a forward location. This will allow you to operate continuously once the attack has begun.

17.9 Many high-rises are designed with large open spaces that are filled with combustibles.

17.7 Just getting firefighters and equipment into the fire will be a major task. *Courtesy of Ron Jeffers.*

17.8 It will take a number of firefighters just to stretch enough hose into a multi-story building. *Courtesy of Harvey Eisner.*

Ventilation

Our third group of concerns centers around the ability of your fire fighting forces to effectively ventilate the fire building. It is essential for you to remember that the insulated style frequently used in curtain walls, not to mention the concrete construction, will hold the heat in and drive it back on your operational teams. Unlike the average building, high-rises have fixed windows that are extremely difficult to open. They are made of materials that cannot be easily broken and their normal operating mechanisms may require special tools that are not readily available. The best time to find this out is during your pre-incident planning expeditions.

Environmental Controls

Our next area of concern lies within the realm of those environmental controls that govern the movement and condition of air within a high-rise building. We have all heard that it may be necessary to operate the building's air handling systems in order to conduct ventilation operations. Unless you can arrange for help in this area, you are in trouble. Remember that even where devices are built in to assist in the movement of smoke, the limited capacity of some HVAC (heating, ventilating, air conditioning) systems can stymie your attempts to make a building safer to operate in.

The first problem lies with the fact that forced air supply mechanisms provide additional oxygen-laden air to the fire area. This can worsen fire conditions and hurt your fire control attempts. One of the things that Incident Commanders must keep uppermost in their minds is that smoke may be recirculated through the very system being used for the infusion of fresh air into the building. Never forget that common ceiling plenums can allow smoke and toxic gases to move well ahead of your operational positions.

Another fact that must be considered is that combustible insulation can burn and lead to a condition known as instantaneous obscuration. This is the sudden dropping of a heavy, toxic smoke cloud from above the ceiling. It makes seeing exits or movement about a floor impossible. And the toxicity of the smoke itself can be fatal. Compounding problem are the myriad ducts, shafts, and connections that can also allow smoke and toxic gases to spread unchecked throughout a floor, series of floors, or a whole building.

The central air conditioning system works as one part of the total environmental control package for a high-rise building. The dangers of these systems are listed as follows:

- Forced air supply can fan the flames

- Easily recirculates smoke

- Insulation holds heat and products of combustion in and can feed fire.

- Fixed windows designed to maintain an interior environment are hard to ventilate.

- Ducts, shafts, and ceiling plenum problems of smoke, heat and fire spread

The problems that face your fire department as you seek to become proficient in operating within a high-rise environment are many. The variety of ways they can interact are manifold and quite complex. It is only through a planned effort of addressing these matters that your department can have any hope of working safely in these towers of glass and steel.

It's not bad enough that the buildings cause problems. The actual fire fighting operations have a number of problems themselves:

- Large-scale operations: often 4-6 times as many personnel needed to fight fire
- Multi-level operation: lobby control, upper command post, staging, forward staging
- Large area: multi-front operation
- Hostile interior environment: takes its toll on firefighters and equipment.
- Dependence on building systems of all types
- You are forced to consider your sequence of arrival and commitment of forces.

Establishing Command

Let us explore these important concepts in greater depth. The primary command post will be placed at the fire control center, usually found in the lobby area of a high-rise building. From this central control point, the officer in charge of the incident can communicate with those working at each of the other operational points. This is often called the Lobby Control point.

At the fire control center, the Incident Commander will have the best potential to control the building's many operating systems: air handling, fire department communications, elevator monitoring, etc. This primary command post will also be the marshalling point for the other necessary resources such as police, engineering, building, water, and public relations department.

A subordinate commander is further charged with the development and staffing of a forward command post. The forward post is usually one floor below the fire floor. This command position offers a number of operational advantages to the overall fireground operation. The chief at this position is blessed with a good upfront view of the fire fighting operation based upon readily-available intelligence from firefighters who are actually operating only a floor or two above. The exchange between commanders can lead to a better interaction of ideas and operations. This person's major responsibilities are to:

1. Direct the attack based upon experience and available data.
2. Control and coordinate the actions of the attack units.
3. Keep the lobby control command post informed of progress.
4. Account for companies assigned to him.

The Incident Commander has the option of running the show from the forward command post rather than from the Lobby Control point. If that is the IC's choice, another chief officer must be assigned to the fire control center in the lobby.

The key to operational success lies in the communications, coordination, and strategic decision-making abilities of the Incident Commander and key aides. However, it does not hurt to have trained and dedicated firefighters and officers pressing in on the attack.

As we mentioned earlier, the resources and manpower for a high-rise fire are far greater than those required at a normal, ground-level structural fire. To allow for a controlled use of resources, it is important to set up a staging area 2 to 3 floors below the fire fighting operation. This staging area will provide a place to organize your people for the combat role a few floors over their heads. At that area, you will build up your resource pool to include sufficient quantities of the following:

- Manpower
- Air tanks
- Medical support services
- Hose, tools, and additional equipment
- Administrative staff

You will also want to set up a zone within the staging area for the rest and recuperation of your combat forces. It is critical that they have the ability to get their strength back before they rotate back to the fire floor. Refreshments will be a critical part of this process, as food, hydration, and medical assistance are essential components in the process of getting people ready to return to their fire fighting duties.

In order to allow for effective supervision and coordination of your forces, you must consider the number of chief officers that might be necessary for an operation of the size of the average high-rise incident. The following is a level to be considered:

- A sector commander at each command post [lobby and upper floor]
- One at the staging area
- One or more on the fire floor [depending on the number of fronts]
- One or more above the fire

Regardless of the number of sector commanders that your department deems necessary to get the job done, it is essential that they be able to arrive in a timely fashion.

How do you achieve these numbers of chiefs in a smaller departmental scenario? You must look to assign officers from mutual aid towns to some of the subordinate posts. However, if you decide to do this, you must ensure that they then become a part of any future drills conducted by your department. In this way they will know how your organization operates and what their expected roles might be. This use of mutual aid is important for any fire department whose resources are not up to the challenge of a high-rise operation. These fires are ultimately consumers of people. You need lots of help to run a high-rise operation.

Strategy

Once you have decided on a number of staffing-type issues you can turn your thoughts toward which type of strategy to use. Basically there are three strategies available to use at high-rise incidents. They are:

- Direct frontal attack
- Flanking attack

- Defensive attack (actually a holding action)

Direct Frontal Attack. The most commonly used strategy is the direct frontal attack. In this method, the fire officer selects the staircase closest to the fire and has hoselines attached to the nearest standpipe (Figure 17.10). These lines are then advanced directly toward the seat of the fire. Additional lines are moved in to support the initial commitment as needed. The extra lines beef up the attack capability and increase the heat reduction capabilities of the initial attack.

The results from this sort of attack can range from excellent to disastrous. Rules for attack strategy mode selection are not hard and fast. However, there is one instance where the direct attack is mandated: when the means of egress is obstructed by the fire. Under these conditions, the direct thrust of an aggressive attack to the location of the occupants is the only approach that may save a life.

Most initial attacks at high-rise fires begin as frontal attacks. The success of this strategy is usually determined by the size of the fire, the training of the attack personnel, and the aggressiveness and motivation of the attacking force.

17.10 Standpipe operations are the most common fire fighting strategy in high-rises.

Flanking Attack. In the flanking attack, you attempt to head off the fire, or move in on it from two different fronts in the form of a classic pincers operation. You must remain flexible and look at all of the possible ways in which to approach the fire. This can be used when the main body is too hot to move right on in, or when the fire is too big and you must narrow the front.

Defensive Attack. Far too often the defensive attack is overlooked as an early strategic option. Many fire service people are too proud to admit that some situations are too tough for them to handle. In addition, the direct attack usually seems to us to be the easiest and quickest to implement. Let us, however, take a look at some of the signs that can point us in the direction of an early move toward the defensive:

- Limited initial resources (such as the arrival of the only two engines in a very small department)

- Disastrous arrival conditions (heavy fire on one or more floors or perhaps explosions, or both)

Think back to the fire in Sao Paulo, Brazil during the early 1970's where the entire building was burning when the fire department arrived. This is an excellent example of a situation where the defensive mode must be the mechanism of choice. In those cases where there is the need for a defensive attack upon arrival, do not be a hero. You may wish to apply large streams to smash the fire. It is not out of the question to even consider the use of master streams. With strong pressures and large nozzles, a master stream can demolish hung ceilings, destroy the convection currents of the heat-generated fire, and retard the spread of the fire. It is crucial that you work to correctly identify the following facts:

1. Location of fire: floor and area involved

2. Location of all life hazards
   a. Occupants
   b. General public
   c. Firefighters

3. Potential for smoke propagation

4. Actual and potential extension of fire
   a. Vertically
   b. Horizontally

5. Location and progress of hoselines

6. Location and activities of truck companies

7. Reported conditions in all operational sectors

8. Status of building systems

9. Status of your command and control organization

10. Status of communications

At fires in high-rise buildings it is even more important than at regular ground-level structural fires to anticipate your needs and call for help as quickly as necessary. Given all of the problems you would expect at a large-scale high-rise fire, you should make your calls promptly. The following are indicators that additional help is necessary:

- Fire showing on upper floors

- Smoke covering facade of building

- Building fully occupied

- Poorly-staffed first alarm companies

You must design your first-alarm response to cover the higher levels of manpower needed at fires of this type. Normal operational delays in reaching the fire can allow it to get out of hand. You should anticipate these delays and assign additional units automatically. If you lack the resources in your own community, do not hesitate to develop an automatic mutual aid program with neighboring fire departments.

Two final but very important points to remember when attacking high-rise fires are the time factor and the fatigue factor. Any operation that you choose to perform will take longer. The logistics of moving people and equipment is an additional challenge, one that involves time, practice, and planning. The fatigue factor slows your people down because many of the manual fire fighting tasks involve overcoming gravity as well as performing the task. As the Incident Commander, you must consider the extra time and talent needed to perform tactical operations in a high-rise environment.

Life Hazard Considerations

Remember that occupants of the involved building may be exposed to such hazards as:

- Low velocity air movement
- Toxic gases
- Obscuration
- Stair contamination
- Plenums spreading products of combustion
- Elevators carrying them to the fire floor
- People not conditioned to respond to an emergency

Firefighters, of course, are not immune to danger at high- rise fires. They could be taken out of action by such factors as:

- Elevators taking them to the seat of the fire
- Fire encircles and cuts them off
- They could become disoriented
- Ceiling supports could collapse
- They could be knocked out of action by heat and overexertion

It is for reasons such as these that you must plan on at least doubling your personnel requirements. Some authorities call for even higher fire fighting manpower quotas. In any case, such pre-incident planning is critical to your success at high-rise fire incidents.

Elevators

There are a number of elevator-related problems that you can expect to crop up periodically during high-rise operations. They are:

- Call buttons can bring your people to the fire floor
- Safety circuits may not function properly
- Doors can warp
- Door closing systems can malfunction due to electrical or smoke obscuration problems
- You are vulnerable to smoke as an elevator passenger (all firefighters must have on SCBA as they begin ascending to the fire floor).

Your people must be trained to expect these malfunctions and know how to overcome them. Lack of knowledge and training in this area can lead to unnecessary civilian and firefighter casualties.

Tactics

Now we come to an area of extreme importance to the overall success of any high-rise fire fighting operation: the company tactics needed to effectively combat these types of fire-related scenarios. You must ensure that your fire attack teams are thoroughly trained and well-drilled in fighting high-rise fires. In those cases where the fire involves a small area it is essential to use a direct attack on the seat of the fire. The same holds true when you must cover a life hazard on the fire floor. Regardless of strategic needs, saving lives should always be the strongest factor in the placement of a hose stream.

There are going to be those times, however, when you will not have the luxury of being able to use a direct attack. At these times you will have to resort to the tactics of confinement and encirclement. These tactics are best used when:

- The fire is extending rapidly

  The fire is beyond the control of handlines

- Fire threatens the safety of your operating forces

- You face structural failures of the floor, shafts, or exterior walls

- Heavy streams are needed to bring the fire under control

- Fire threatens to encircle and cut off your fire fighting forces

In a confinement and encirclement operation you seek to mount a sufficient number of hoselines to be able to ring the fire area with water. You then must work to limit its ability to continue its destructive onslaught. This may well require the use of multiple standpipe outlets, each of which has multiple lines operating through a gated-wye device. Do not even think of attempting this if you have not devoted time to it on the drill ground. You will be jeopardizing your people on a task for which they have not been trained.

Safety Operating Guidelines

It is of the utmost importance that your fire attack teams strictly adhere to the following standard instructions for operating at high-rise fires. Failure to follow these guidelines can have catastrophic outcomes. They are:

- Do not overload elevators (observe posted weight limits).

- Be sure that each person has SCBA, tools, or a length of hose (no one is to go up to the fire floor empty-handed).

- An occupancy limit of 5 to 6 firefighters per elevator is suggested as being appropriate.

- Each elevator load should have at least one forcible entry tool among its equipment complement.

- Each unit commander must be sure to report company designation, car number, and destination to the command post before entering the building for operations. Personnel accountability is a critical life safety factor in high-rise fire fighting operations (Figure 17.11).

- As you move into an elevator, determine its relationship to the stairs, just in case you have to make a sudden, unplanned exit from a stalled elevator.

- If possible, use only those elevators that can be placed under the control of fire department personnel.

- No matter who you are, always report in at the command post upon arrival (both lobby control and forward command post when used).

By observing these basic operating rules, you can minimize any unnecessary exposure to injury and death.

Equipment

No matter how well trained your people are, your operation will fail if you do not take steps to ensure that your people have enough equipment. Be sure to give thought to each of the following:

- Hose and nozzles
- SCBA
- Extra air supplies
- Tools (forcible entry and wrenches for use on standpipe risers)
- Adapters for use on standpipe risers (gated wyes, gate valves, etc.)
- Portable radios capable of working at a high-rise location (correct type and sufficient numbers)
- EMS coverage for injured civilian and fire personnel
- Rehab facilities for fire personnel

17.11 Keeping track of everyone at a high-rise incident will be a major job by itself. *Courtesy of Harvey Eisner.*

In planning for any potential fireground operations in high-rise buildings, there are a number of considerations that will apply, just as they do to normal ground-level operations. Of primary importance is the fact that you may have limited manpower in the early stages of your operation. Do not allow the first few units to become overcommitted in the absence of sufficient backup people and supplies. All operating personnel are susceptible to the effects of heat, just as they would be at a fire in a normal 3-story frame dwelling. However, they may be more tired from walking up stairs, or the heat may be greater because the building holds it in. Guard against overaggressive operations under conditions of minimum staffing availability.

Be sure to provide adequate levels of supervision, command, and control. Your operational framework must provide for reporting procedures to ensure that you do not spend your scarce manpower resources unwisely. In order to do this, you should plan to use hardwire fire department phones whenever available. And always maintain an organized and disciplined approach to communications. (One screamer can ruin your operation and spread panic and confusion.)

During your planning operations, remember that each high-rise building is just a little bit different from the next. You must be aware of these variations. It is important to work with building management to see that occupants and service employees know what to do during an emergency. Basic staff preparation is essential. Be aware of such basic design features of high-rise in your community as the following:

- Core design

- Building components

- Exit arrangements and areas of refuge

- Fire control operating systems

- Firefighter's phone jack locations

- Operational aspects of building system components such as heating, ventilating, and air conditioning.

The department should know the location and operating procedures for emergency power operations in a high-rise. Elevator controls and their operation form a very basic part of your fire attack. Know where to gain control of them and how to safely use them. You should also be aware of the various floor plans for the high-rise structures in your area. Where possible, secure copies of these floor plans for your operating forces and conduct both blackboard and on-site drills in each facility. During a working fire is not the time to discover that you are a stranger in your favorite local high-rise building. Remember that the keys to efficient high-rise fire fighting are:

- Knowledge

- Training

- Planning

You must know what to do, deliver training to allow your people to do it, and above all, plan to bring it all together.

CHURCHES AND OTHER RELIGIOUS OCCUPANCIES

It has been my experience that church fires are usually a losing proposition (Figure 17.12 a and b). They are generally so big and so complicated that fires can build up a good head of steam before they are found. They may be quite old or quite modern. In this section, I am not referring to the smaller store-front type of churches found in many cities; those can usually be handled like a neighborhood commercial store. I have attended a number of these and that was our general operational mode. Larger churches are a creation all by themselves. When you add to this the fact that most older religious facilities, and many modern ones, are built primarily of wood, you have the potential for a spectacular blaze. The ones I have attended over the years have been all of this and more.

Problems:
- Occupancy loading is uneven. There may be no one in a church.

- A great many concealed spaces

- Steeply pitched roofs (or at the very least, really ornate ones)

- Often limited access from the rear

- High ceilings (Guess where the cathedral style of ceiling came from)

- Large, open areas with minimal division from the attached administrative areas

- A steeple can serve as a flue

- Exposures can be placed in danger because of their closeness.

- It is tough to make an inside attack in a well-advanced church fire.

- Limited number of entry points into a large building

- Tough to use heavy streams because of limited number of windows and doors.

- Balconies and heavy lighting systems can fall onto fire attack forces.

Let me reinforce my first comments about the minimal possibilities of successfully controlling a church fire. My first multiple-alarm fire as a chief was in a historic, downtown church. When we arrived, smoke was coming out of every place available. By the time we were able to force entry into the locked building, the fire had spread from the main body of the church into the attached school area.

The primary battle involved exposure control. The use of master streams and 2-1/2-inch handlines with solid-bore tips protected the apartment house next door. Handlines were also stretched up the interior stairs of the exposure building. We then used them to attack the main body of fire through the holes that had burned in the roof. Needless to say, the place was a total loss.

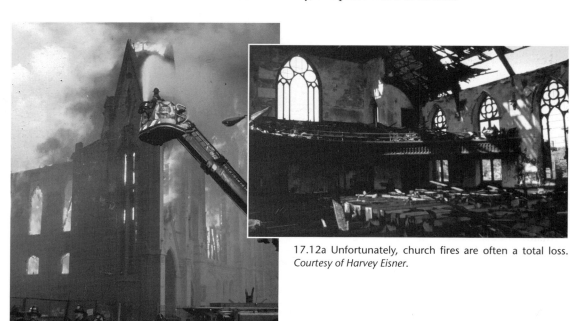

17.12a Unfortunately, church fires are often a total loss. *Courtesy of Harvey Eisner.*

17.12b The large, open spaces inside churches contribute to rapid fire spread.

On another occasion we were able to save an adjoining church school. By moving 2-½-inch handlines in through side doors, we were able to drive the fire back into the church and away from the doorways into the school area. This was not an easy operation because we were dealing with a 75-year-old wood frame church. Luckily, firefighters were able to ventilate the roof by operating from ground ladders. The roof was only slightly peaked. This was most fortunate because church roofs usually have a high pitch, and others are covered with such old-time niceties as slate roofs and copper drains and downspouts.

There are some religious artifacts that you should save, if the opportunity presents itself to do so in a safe manner. Saving historic copies of the Bible, the Torah, or the Koran, for example, can help a congregation cope with the loss of their home. However, the lives of your firefighters should not be jeopardized. Proper salvage practices can also go a long way toward limiting damage. Be very careful in fire scenarios involving churches. Do not let your urge to help overcome your instincts to operate safely.

LUMBERYARDS

If ever the heart-warming college homecoming phrase BONFIRE had an apt real-life counterpart in our world of fire fighting, it is the lumberyard fire (Figure 17.13). These can be truly spectacular challenges to your exposure control abilities. I have only attended one, but its memory is quite vivid. Our success in that case was due to an excellent water supply system and a large fire department. The three things you will need to control a lumberyard fire are:

17.13 Controlling a fire in a lumberyard, with its enormous fire load, will take huge amounts of water and personnel.

People • Water • Luck

Problems:
- Large volume of wood (Hence the name LUMBERyard)

- Outside storage close to inside storage and woodworking areas

- Strong potential for downwind brand and ember problems

- Collapse hazard from unstable piles of lumber

- Overhaul can become a problem if fire burrows deeply into piles of stacked lumber

- Delivery vehicles can easily become involved in fire if stored close to lumber.

- Access to all parts of facility may be difficult.

As I said earlier, the keys to success in combating a lumberyard fire are people, water, and luck. You will need to deploy large streams in an effort to muscle the fire under control. If you do not have the people to staff the hoselines or the water to supply them, your best chance of success is to aggressively cover exposures with the available fire fighting water. Where does the luck come in? In a number of ways:

- Wind blows fires away from exposure structures.

- Fire occurs during a time when your personnel response is adequate.

- Alarms work properly and detect fire early.

- The indoor facilities are protected by an automatic sprinkler system.

After you have been successful in knocking down the heavy body of fire, you will still be faced with an extremely dangerous task. Overhaul in these occupancies should be undertaken cautiously and carefully. Do not commit personnel to areas where stacks or structures are in danger of collapsing. As always, it is important to operate in a systematic manner. Do not let the front-end loader people or the forklift operators move lumber around in a random fashion. You should also avoid scattering lumber as the piles are broken up. You may accidentally hide smoldering lumber. This can come back to bite you later in the form of a rekindle. Here are some other concerns you may be called upon to address:

- Avoid shooting water at each other from opposite ends of lumber piles.

- Watch for falling electric wires that have been damaged by the high heat of the fire.

- Check the area surrounding the fire for brands and embers. Have additional units respond to perform this task.

- Always cover exposures. If possible, stretch hoselines into nearby buildings.

- Remember that there is a back to the fire. It may be hard to reach through security fences and the like, but you have to ensure that the fire is seen from all sides.

- Avoid overhauling at night, if at all possible. If you must do so, provide lots of lighting.

- Be prepared to keep a standby crew on location for a long time, days if necessary. They should maintain an established water source and hoselines ready for use.

- Do not forget that there may be a propane storage area for cylinders used by forklifts.

Fires in lumberyards are a challenging event, regardless of the size of your fire department. Pre-incident planning is a critical element in your preparations. The more you know, the better you should do. Should you happen to encounter one, keep in mind that people should never be placed in jeopardy for the sake of property. Do what you can with the resources available. Call more as needed. But be careful.

SCHOOLS

No one likes to fight a fire in an educational occupancy. If the building is occupied, a lot of lives are at risk. And even if no one is there, much harm can come to an educational system if a school is lost (Figure 17.14). I have attended a number of these incidents. In each case we were most fortunate to be part of a team that saved the structure. The kids were evacuated and the spread of the fire halted. We were most fortunate. However, my experience involves a number of daytime incidents. In these cases, the fire was quickly located and promptly reported. Combating fires that have started at night has not always been so successful.

17.14 The life safety hazards in a school fire will vary widely depending on the time of the incident.

Problems:

- Life safety hazard during occupied periods. There may be confusion over just who is in school on a given day.

- Children may not behave rationally. They may not know how to react.

- Chemical labs at certain schools

- In older schools you may have problems with fire entering the ductwork and spreading unimpeded throughout the building.

Each action you take at a school fire should be tied to the concept of life safety. Search and rescue are your primary concern. Ventilation operations should work to channel smoke away from where the students are evacuating. Hoselines should be positioned to cut off the spread of fire. Command and control may become confused in the rush to do something about evacuating the students. Your command system will be sorely tested at these incidents.

Visits made for public education can pay great dividends. Students will know about fire safety and evacuation, and you will come to know the students and the schools.

WAREHOUSES AND INDUSTRIAL FACILITIES

It is difficult to pin down exactly what we mean when we speak of warehouses and factories. To some, images of large brick buildings come to mind. To others, expansive modern buildings are the norm. The important thing for you to do is learn about the hazards from factories, warehouses, and industrial facilities in your community (Figure 17.15 a and b).

Problems:

- Large old brick and wood-joisted structures can burn for days.

- Industrial processes may give off toxic fumes.

- Older buildings have open stairways and many process openings between the floors, or through walls.

- Heavy water flows may be needed.

- High heat will force an outside fire fighting operation.

- Water supply in older industrial neighborhoods may be substandard.

- You may encounter a windowless structure where ventilation is difficult, if not impossible.

17.15b Newer warehouses are likely to built with lightweight construction.

17.15a This old urban warehouse presents a significant challenge because of its contents and exposure potential. *Courtesy of Harvey Eisner.*

- Movement into building may be hindered by limited access.

- Movement inside of warehouse may be blocked by stock.

- Material may be stacked too close to adjacent piles or in stacks that are too tall. They may fall and crush fire personnel.

- Modern lightweight roof and ceiling assemblies will not support fire personnel. Topside ventilation may not be an option.

- Your needs for forcible entry will vary by time of day.

The key to success in an industrial or warehouse fire may be the amount of time you spend pre-incident planning the building. While there may be certain general characteristics among these buildings, there are a great many site-specific matters. You will need to know as much as possible about such things as:

- Processes involved

- Materials made

- Materials stored

- Number of people working on each shift

- Size of office staff

- Availability and condition of automatic sprinklers, standpipe systems, and alarm systems.

- Location and condition of firewalls, fire doors, and draft curtains

By being aware of such information, you can make better decisions under emergency conditions. When you know what is available to help you, where it is located, and if it is working, you can do a better job of controlling an industrial or warehouse fire.

SUPERMARKET FIRES

The tremendous outward migration of people from our cities over the past 50 years has spawned a type of occupancy familiar to us all. What might have been a neighborhood store in the city has expanded into an all-service shopping facility in the suburbs (Figure 17.16). While a great many of these are tied into strip malls, others are stand-alone operations. It is possible to meet a great many needs with a single shopping stop. It is this range of goods and services that leads to problems during fires involving supermarkets.

Problems:

- Limited movement space because of aisle layout

- Heavy combustible loading, both on shelves and in storage

- Access from store is limited by checkout spaces. This leads to limited exit availability.

- Occupant loading varies greatly by time of day and time of week.

- Large open space necessitated by selling environment

- Lightweight truss-style roofs or metal deck roofs make rooftop ventilation a questionable alternative.

17.16 Supermarkets will challenge you with access problems, exposure problems, and fire load problems. *Courtesy of the Phoeniz (AZ) Fire Department.*

- When steel bar joists are used as support mechanisms, they can absorb heat, expand, and cause walls to kick out and collapse.

- Space above hung ceilings can allow fire to travel unseen by occupants and fire forces alike.

- When built as part of a row of stores, openings may exist between buildings. This can serve as a route of rapid fire travel.

- Hose streams may encounter a wide range of obstructions.

- Storage may be at the rear of the main store area, in a basement, elevated above the store, or some combination of these methods. The only way to know is to visit, look, and learn.

- Automatic doors may fail, trapping people in the store. This can apply to ANY STORE, anywhere, so always have it on your mind.

Conducting fire fighting operations when a busy supermarket is open can be very difficult. Rescue is your primary function. Do those things that will save the greatest number of lives, while taking care to limit the danger to your crews.

Access may be limited by both legal and illegal parking. Hydrants may be blocked and hose stretches difficult for the same reason. It may be difficult to get pumpers to the front of the building through crowds of evacuees. Care should be used when maneuvering equipment in for the attack. You will be forced to move your hoselines in through the same doors through which people are fighting to exit. One of the classic tactics you may wish to quickly consider is to remove the large plate glass windows at the front of the store. If done carefully by trained personnel, you can remove a great deal of smoke. And you can offer an alternate way in and out of the store.

You may be forced to direct hose streams over the heads of fleeing patrons. While this is tricky to coordinate, it can be a lifesaving tactic. Remember to use multiple hoselines on the attack. And always back up a hoseline with one of equal or greater capacity. Frequently this second line will be needed to absorb enough heat to allow progress by the initial hoseline crew. I have also seen the second line protect the evacuation of the first.

It is critical that your personnel check for extension early in the operation. Whether it is internal or external, the spread of fire can make your job more difficult and dangerous. And it can spread the damage if you are not careful. Some of the truck company tasks that you will need to consider are:

- Forcible entry
- Ventilation (consider positive pressure if available)
- Place ladders as circumstances dictate. Be careful of rooftop operations. Firefighters have been killed when roofs on supermarkets have collapsed.
- Open up walls and ceilings to check for fire spread.

Once again, these are the sort of fires that respond to the efforts of a lot of people throwing a great deal of water. You should consider automatic aid if you have insufficient forces to handle a fire in an occupancy of this type. A few extra people rolling immediately are much better than scores of help that arrives too late to do any good.

PETROLEUM STORAGE FACILITIES

Regardless of where you live, the potential exists for you to be called to a fire in a petroleum storage facility. Whether it is a local fuel oil delivery firm or a major regional storage facility, the challenges are similar (Figure 17.17).

This book does not cover operations at a major refinery incident or a major tank farm fire. That is a separate, special type of fire fighting response, and is beyond the scope of this text. This section is designed to help you handle the problem down at Joe's Fuel Oil Company.

Problems:
- The materials can explode and shower your personnel with burning petroleum product.
- The vapors from many flammable materials can travel low to the ground and be ignited accidentally by outside sources.

- Other vapors will rise and move away, causing an off-site explosion hazard.

- Surrounding property is subject to tremendous radiant heat and may quickly ignite and spread the fire.

- Delivery vehicles with product onboard may be stored inside a building. This can be a very bad situation in terms of fire load.

- The return to aboveground storage for environmental reasons places the problem squarely in your face.

Petroleum products are designed to operate within a closed system. They are not supposed to see the light of day. From their beginnings in an underground pocket of oil, through the refining, transport, and distribution stages, they are supposed to be within a system. Our problems come when something causes that system to fail.

Experience has led me to form a number of rules for operating at incidents involving flammable and combustible liquids (Figure 17.18). I urge you to adopt the following and train on them:

- Use coordinated attack teams. Multiple units can protect each other. At one flammable liquids fire, I was saved from burning up by a backup foam line.

- Use the proper agent for the material involved. No one foam will work on every burning material.

- Do not ignore the exposures to fight the fire. You need to consider additional protective hose streams for areas exposed by burning flammables. This rule holds true whether you are dealing with vehicles on the highway or a storage situation.

17.17 Even a small petroleum delivery trucks will be a major hazard.

17.18 Use the right extinguishing agent for the product involved. *Courtesy of Harvey Eisner.*

- Keep the troops upwind.

- Work from higher ground. In this way, the burning fuel will not run down the grade and attack your operating personnel.

- If at all possible, stop the flow of fuel. Sooner or later, no fuel means no fire.

- Surround the fire with fire streams.

- Work to develop sufficient flow to absorb the heat being given off by the fire. It may take a great deal of water, time, and talent.

- If possible, have facility personnel shift petroleum products from the bottom of the burning tank to other tanks that are not in danger. Do not let your people perform this critical task.

- Have potentially exposed delivery vehicles moved as quickly as is practical.

- **DO NOT GET TOO CLOSE TO BURNING TANKS, BECAUSE THE PRODUCTS CAN LEAVE THE CONTAINER THEY ARE IN AND BURN YOU.**

Eventually the fire will go out; the object is to achieve this result without injuring your people. You must consider that this type of fire could happen in your community. Train your staff. Provide resources for their use. Seek help from the owners of the properties in question. Be prepared.

AIRPLANE CRASHES

No matter where you live, it is hard to envision a place where a plane crash could not occur. Air travel is such a great part of life that an aircraft can plummet from the sky right into your backyard. It could be a major public carrier, a commuter aircraft, a military plane, or a privately owned plane. It is entirely possible that your local fire department may be the only fire protection service available for a small local airfield.

I can remember the night when a small military helicopter skimmed over our home at a terribly low level. As I commented to our barbecue guests that the plane was well under flight limits for our area, my volunteer fire pager started beeping. We were alerted to the crash of a light observation chopper into a local blueberry field. Fortunately there was no fire and the injuries were minimal. The pilot and observer were quite lucky. Such is not always the case.

Problems:
- Aircraft carry flammable liquids on board to power their flight. These liquids can be explosively released on impact.

- Modern airliners can carry over 300 passengers.

- Cargo planes carry a wide variety of flammable, combustible, and hazardous materials.

- The operating systems for aircraft use flammable and combustible materials.

- The metal frame and body of the aircraft can burn intensely.

- Falling planes have no respect for property lines. When they fall, they just fall. And if they land in your town, tough on you.

- Military aircraft carry a wide range of ordnance, ranging from missiles, to bombs, to machine gun and cannon ammunition. Approach a downed military aircraft from the rear, if you must approach at all.

The results of a plane crash can range from minimal damage, to a plane landing under a pilot's control, to total devastation when they plow into the ground. Your ability to influence any potential rescue and fire fighting outcomes is tied directly to the impact damage of the crash itself. The two major types of aircraft incident are:

- **High impact** (Devastation)

- **Low impact** (Walk away from it)

Either type can lead to a very hot fire (Figure 17.19). You will need to be able to apply a great deal of fire fighting foam to handle these incidents. We will make the assumption that your ability to apply fire fighting foam is limited to one of the following:

- Handline eduction and application

- Around-the-pump proportioning and handline application

- Delivery of around-the-pump foam through a pre-piped master stream on a pumper

- Delivery of around-the-pump-generated foam through a hoseline to the piping of an aerial device

You should become proficient in the use of these methods (Figure 17.20). An aircraft incident is not the place for on-the-job training. You will need to deploy your forces quickly and decisively. You will need to have a pre-incident plan in place to handle this type of emergency. Your plan should include the following:

17.19 There are often survivors to tend to, as well as fire, at a low-impact crash. *Courtesy of Joel Woods.*

17.20 Your training operations will need to include the application of foam in order to handle these situations effectively. *Courtesy of Joel Woods.*

- Types of aircraft flying in your area
- Potential crash sites
- Flight patterns in your area
- Aircraft fire fighting and rescue techniques
- Periodic rehearsals of the plan
- Update as needed

In this type of incident, it is critical to limit access to the site. If there are fatalities, you will not want people wandering around. They may disturb evidence or steal possessions of the passengers. Should there be fatalities, treat them with dignity, but do not move them until the proper authorities have responded to begin their investigation. There is no need for moving the deceased.

In the event of survivors, Emergency Medical Service units become a priority. Your pre-incident plan should, therefore, include the response of EMS. While their skills are separate from fire fighting, their operations should dovetail with yours during this type of emergency response. And once the incident is terminated, keep the feelings of your people in mind. There may be a great need for critical incident stress debriefing teams to interact with your people. This element should also be a part of your plan.

If the aircraft involved is military, be sure that the nearest facility is notified. There are protocols that the military must follow. You can speed their arrival if they are notified in a timely fashion. They can be of assistance in situations involving exposure to exotic weapons or materials. The military can also be a source of assistance in your pre-incident planning.

Summary

This chapter has covered a number of large and diverse incidents. Some of these events might might strike the average fire department once in a generation.

My goal in describing them is to leave you with a single, simple thought: **No matter what you face, the need to maintain a consistent focus for your operation remains the same**. If you learn the eight steps to sizeup, and practice them faithfully at every incident, they will guide you in the right direction, whatever catastrophe you and your agency are called to combat.

Remember that sheer size is not a good reason to panic. In fact, while I was preparing to write this summary, I was called to a truly once-in-a-lifetime fire. I was the first-due Battalion Chief at an incident involving a blazing high-level harbor crane. We had heavy fire involving the following:

- Diesel fuel
- Hydraulic fuel
- Hydraulic pumps
- Electric motors
- A wide array of combustibles

We also had to deal with a fire that was 40 feet above the ground and an exposure problem presented by a large, ocean-going ship. Fortunately, the crane operator self-evacuated as our first units were arriving. I have been to harbor crane collapses in my long fire service career, but never a working fire.

However, through the use of my eight-step process, and some excellent fire fighting by our crews, the blaze was rapidly brought under control through the use of four elevated streams. When the time was right, we moved in with foam hand lines and mopped it up.

The system worked for me, and it can work for you.

18

Construction Issues

In order for you to develop a thorough knowledge of the field of fire fighting strategy and tactics, you must be familiar with building construction and its relation to fire.

It is important for fire personnel to understand just how the various components of a building fit together into a functional building system. Let me use an analogy from our friends at the International Fire Service Training Association (IFSTA) to give you a proper feeling for just what we mean. If a building is viewed as a system (much as the human body is a system), then a fire inside is an attack upon the system. This happens in much the same way that infection attacks a living body. As in the case of a living body, the course of the fire will be greatly influenced by the way in which the building reacts to it. However, unlike the bodily system referred to above, the building has to contend with such things as people and contents, things that can make a good system bad and a bad system worse.

Learn about the major types of building construction so that you will know what they can do to help or hurt your fire department operational procedures. Your life and the lives of your firefighters may depend upon your skill and judgment. You need to gain a proper appreciation for all that goes into something as simple as the word (and concept) *building*.

A caution: what follows is only the beginning of what you need to know. It would be my suggestion that one other book needs to be added to your library in order to build upon these rudimentary comments: IFSTA's **Building Construction Related to the Fire Service**. This book is a reference source sufficient to handle any normal building construction-related questions you might encounter.

TYPES OF CONSTRUCTION

There are five basic types of building construction currently in standard use throughout North America. They are listed below:

- Type I (Fire Resistive)
- Type II (Noncombustible)
- Type III (Exterior Protected Combustible)
- Type IV (Heavy Timber)
- Type V (Wood Frame)

There are also a wide range of buildings that can be classified as being of mixed construction. This happens when parts of two or more different types of construction are used in the same structure. Every person who responds to fire fighting operations must know the various building types and what makes them so important.

Type I Construction (Fire-Resistive Construction)

In fire-resistive construction, the structural members are noncombustible and are fire protected (Figure 18.1). In these buildings, materials are used that increase the existing noncombustible rating. It is important to stress that for a construction type to be rated as "fire resistive," only noncombustible building materials can be used for the structure's support elements.

Height and area requirements form the basis for many elements of the model building codes. These are based upon a strict interpretation of the definitions for each of the building types. Compromising the integrity of any particular subpart of the building would endanger the total rating of the building. A good example of this type of mismatch would involve the use of unprotected wooden interior bearing walls in a Type I, fire-resistive building. Given the trade-offs for fire-resistive construction, any lessening of the standards holds the potential for future tragedies. These would become the weak links in your chain of structural integrity.

Type II Construction

Type II construction was once called noncombustible. In Type II construction, the structural members are of noncombustible materials. In this way, fire is not allowed to spread throughout the building via concealed spaces that are filled with combustible materials. Further, the various model codes prescribe various height limits for this construction style (Figure 18.2).

Type III Construction

In the Type III construction scenario, all materials that form the interior makeup of the structure can be of a combustible nature. This is allowed so long as the fire resistance requirements of the appropriate model code are complied with. The exterior bearing walls must be of a noncombustible or limited combustible material (Figure 18.3).

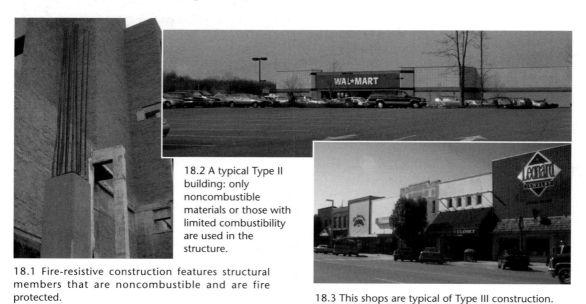

18.2 A typical Type II building: only noncombustible materials or those with limited combustibility are used in the structure.

18.1 Fire-resistive construction features structural members that are noncombustible and are fire protected.

18.3 This shops are typical of Type III construction.

Type III construction is further broken down depending on whether or not a one-hour protective rating is provided for the structural elements. This is of obvious importance to suppression troops. This rating can buy you some very critical extra moments in which to perform search and rescue or to get that quick attack line into place. That is why it is so critical for your personnel to be sure that the building is built and maintained according to your local building and fire prevention codes.

Type IV Construction

Type IV is found in a great many larger, older cities in North America. In this type of construction, the structural elements are usually unprotected wood with large dimensions. Another important aspect of Type IV construction revolves around the fact that with minor exemptions, concealed spaces are not permitted in the floors and roofs or other structural members. Another name for this type of construction is mill construction (Figure 18.4).

To provide the structural stability for this type of construction, minimum nominal widths for various structural elements are specified in the various model building codes. One important feature of this type of construction is that it has a better resistance to structural failure than a conventional wood frame building. This is because the structural members have a smaller surface-to-volume ratio and take longer to burn.

Type V Construction

In Type V, wood-frame construction, the entire structure can be built from wood or any other material allowed by the various model fire codes.

18.4 Type IV, heavy timber construction, may be seen in large, older structures.

This type of construction is of particular importance to firefighters for two reasons: (1) The wood-frame building is the one you are most likely to encounter in your community; (2) This type of structure poses the greatest problems to the firefighter in both interior fire fighting and potential spread to other structures (Figure 18.5 a and b).

Wood-frame construction is probably more vulnerable to fire than any of the other types of construction. Even when properly built, there are a number of open areas within the framework of a wood-frame structure that allow fire to grow and spread. Like Type III construction, Type V construction is divided into two subgroupings. These are based upon whether or not the structural members and exterior walls have a one-hour fire resistance rating.

There are two basic types of wood-frame structures: the balloon frame and the platform frame (Figure 18.6):

18.5 Type V, wood frame construction, is probably the one you will see most often. It is also the one most likely to catch fire.

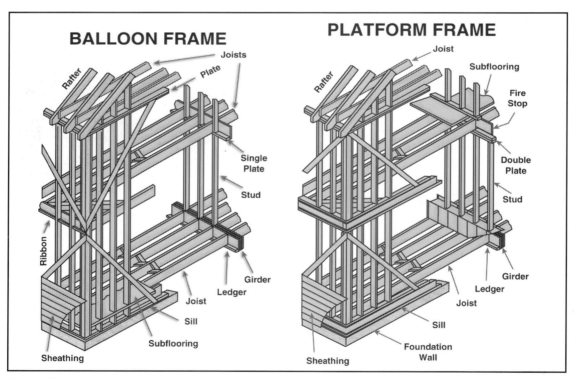

BALLOON FRAME

Rafter

Joists

Plate

Single Plate

Stud

Ribbon

Girder

Ledger

Joist

Sill

Subflooring

Sheathing

PLATFORM FRAME

Rafter

Joist

Subflooring

Fire Stop

Double Plate

Stud

Girder

Ledger

Joist

Sill

Foundation Wall

Sheathing

18.6 Platform-frame construction decreases the number of paths that enable vertical fire spread. Older balloon-frame buildings have open paths that can extend up the entire building.

Platform Frame. In the platform frame, each floor is built on its own platform. This method lowers the number of open paths through which fire can spread upward. The fire service has long been in the forefront of working to require the use of platform construction to make the homes of America safer for us all. However, there can still be gaps and cracks in the code through which fire can sneak. I can recall a structural fire a few years back where the builders had slipped one over on all of us. The building in question was located in one of the newest housing developments in Central New Jersey. The fire started in the basement and spread upward through the cold air return ducts for the home air handling system. The building had been built to the requirements of New Jersey's *Uniform Construction Code.* It was a platform-frame dwelling. Unfortunately, the effect of the cold air ducts that ran under the floors and above the ceilings was devastating.

The spread of the fire through that development home was as near a duplication of a balloon-frame fire as we ever attended in the older sections of Newark, New Jersey. Additional pre-incident planning taught us that there were a number of these ticking time bombs waiting for us. We planned accordingly and will respond to any fire in that development as though it were a balloon-frame dwelling. Moral of the story: Just because a building may be a platform frame does not mean that architects and builders cannot create alterations that return us to the days of yesteryear. Pre-incident plan carefully and be ready.

Balloon Frame. Balloon-frame construction belongs to an earlier era. It was found to be quicker and less expensive to have the structural members run upward from the foundation without any stops, such as are found in platform frame construction. Balloon-frame buildings can, therefore, have open channels that run from the foundation to the attic. There is nothing built into the wall to stop fire from racing upward.

18.7 Get out in your community regularly and see how buildings are being put together. This knowledge can save lives when you must decide what type of attack to mount on the

There is only one way to become familiar with the types of construction in your community. You must get out and visit as many structures as you can during each shift you work. It is important to note that you must combine your visits with sufficient reading sessions. Compare what you find in your towns with what the books say about construction types. This area of the fire protection world serves as a classic example of an old learning theory. You must read about a topic, look for examples of that topic in the real world, and then adjust your textbook knowledge based upon the reality you find (Figure 18.7).

HOW BUILDINGS ARE BUILT

If we are to operate effectively in burning buildings, it is important that we know how they are built. While the materials can vary greatly, the manner in which they are put together and the elements that hold up a building remain fairly constant. Let us begin at the very beginning, in terms of what it takes to put a building together and make it stay up.

The fundamental structural support of a building is normally afforded in one of two ways. A structural frame can be used that is designed to support the applied live and dead loads. Or it could be that the walls are used to support the loads. Both of these approaches are widely used.

Loads

The term *load* refers to the physical forces that are brought to bear on the members of the building and on the building itself. These types of forces are listed as follows:

- Dead Load – the weight of the building and all fixed elements (Figure 18.8)

- Live Load – movable elements such as people and furniture (Figure 18.9)

- Static Load – slowly applied and nearly constant, like the filling of a water tank

- Impact – short term, immediate (like a firefighter jumping from a ladder onto a roof)

- Repeated – intermittently applied like a rolling bridge crane

There are also two other styles of loads that must form a part of the Incident Commander's tactical response review. These are the

- Concentrated load – weight is delivered over a small contact area, thereby concentrating the load.

- Uniformly distributed load – weight that is constant over an area.

It is critical for you to understand just how loads are applied to a building. Loads are basically classified according to the direction that they apply to structural members. Axial loads are applied along the structural member's axis. Eccentric loads are applied to one side of the cross-section of a structural member, creating a bending tendency. Torsional loads produce a twisting effect that creates shear stresses in a material. How a load is placed in a building can have an impact on how the building will respond to a fire situation (Figure 18.10).

The response of the architect to the problem of loads comes through building design. The structure is designed in such a way that it is either strong enough to support the downward thrust of the weight of a particular load or resilient enough to stand the thrust of a load that is delivered from the side. This requires a knowledge of the physics of loads as well as an understanding of the relative strength of the range of building materials. Any discussion of force must consider the effects of structural engineering to the firefighter.

Load-Bearing Walls

When the weight of the various loads is actually carried by the walls themselves, we term these "load-bearing" walls. In buildings of brick-wood joist or masonry construction, great use is made of load-bearing walls. During your inspections of these properties, you must determine whether the interior walls are load bearing or not. A good way to find out is to check the plans for the building, if they are available. Another way to determine this fact is to look for the use of columns. If there are columns to carry the weight, chances are that the walls are only acting as partitions to divide the interior of the structure into usable areas. However, the best way to confirm this is to review the building plans.

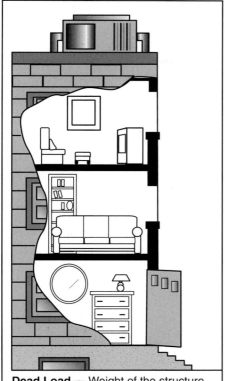

Dead Load — Weight of the structure itself and any equipment permanently attached or built in. Concrete, steel, wood, air conditioning, plumbing, etc.

18.8 The term dead load refers to the weight of the permanent part of the building – the floors, walls, supporting columns and the like.

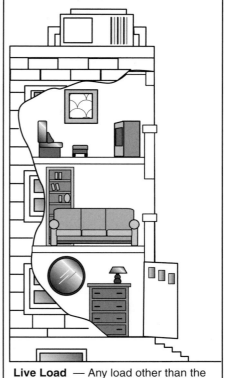

Live Load — Any load other than the dead load. Furnishings, people, personal property, etc.

Note: Water for **fire fighting is a live load**, which is not calculated in the design of a building.

18.9 Live loads refer to what goes into the building – furniture, equipment, and people.

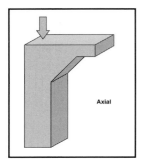

Axial

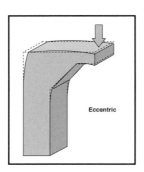

Eccentric

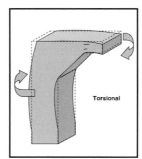

Torsional

18.10 The way in which loads are applied can affect the way the building responds to fire fighting operations.

Where the structural frame is used for primary support, the exterior walls serve only as the structural enclosure, and are termed curtain or panel walls. Many fire-resistive and noncombustible buildings have an interior steel frame, to which are attached floors made from a variety of construction materials (Figure 18.11).

The simplest exterior wall to build is the masonry, load-bearing wall. Whether it comes from block, brick, or stone, this type of wall provides excellent support and outstanding resistance to the ravages of fire. Masonry's versatility has led it to be used in a wide range of roles over the years, such as:

- Exterior load-bearing
- Interior load-bearing
- Fire walls
- Partitions
- Fire partitions

Core Construction

A common style of interior support frame is found in high-rise buildings. A metal frame extends outward from a central core, much like the branches of an artificial Christmas tree. The walls serve to keep out the elements of nature, not much more. Real problems can occur if the fire stops that seal the open spaces between the floors and the exterior panel walls are poorly installed or missing. One need only look to the Interstate Bank Fire in Los Angeles a couple of years ago to see how bad this problem can be (Figure 18.12).

Fire Resistance

Fire resistance is a collective property of materials and assemblies. Basically, fire resistance is the ability of a structural assembly to maintain its load-bearing ability under fire conditions. When we speak of walls, floors, and ceilings, we usually refer to the ability of these to act as a barrier against fire spread. The materials that are used in a particular structure must meet certain fire resistance rating requirements, which are based on a number of factors. These include the size of the building and the overall type of construction that has been approved for the facility. Remember that, although no material is exempt from damage by severe and continuing exposure to fire temperatures, it is necessary to specify the desired fire resistance of materials to be used.

18.11 Steel framing is common in fire-resistive and noncombustible buildings.

18.12 Fire can spread rapidly through the large, open spaces that are common in high-rises. *Courtesy of Los Angeles City Fire Department.*

Basically, there are four elements to the fire resistance equation:

- Combustibility - will the material burn, and if so, how fast?
- Thermal conductivity - will fire spread through the material?
- Chemical composition of a material
- Dimensions of the material, as its shape and form allow for spread

Compartmentation

In discussing just what stresses or problems a fire might place upon a building and its systems, it is important to note that heat itself is a tremendous force and its destructive effects are legendary. The ability of a fire to spread is directly related to any building layout and tightness of construction. If the building is not compartmentalized and fire occurs, it will rapidly spread to destroy the entire building. Further, even a compartmentalized building will not stop the spread of fire if the construction methods are poor and the building is not tightly built (Figure 18.13).

18.13 Learn as much as you can about layout and compartmentation in the structures to which you may be called.

The opposite situation is of great importance to fire fighting forces. A well-built, compartmentalized structure will limit the spread of fire and buy the fire department time to arrive and begin suppression operations. A fire officer who knows how a building is compartmentalized can use these facts to help the operation.

I can recall a fire where my volunteer department was dispatched to a reported structural fire during the early morning hours. Upon arrival it was determined that we faced a cellar fire, so we proceeded to attack it accordingly. After the fire was extinguished and entry made to all areas of the building, we made an interesting discovery. The fire had started in the basement, built to a flashover stage, flashed over, and then died out for lack of enough oxygen to sustain combustion. The building was both well-compartmentalized and extremely well built. Although damage was extensive in the rear of the basement, the fire did not spread to either the front of the basement or the first floor. This greatly helped our fire fighting operation.

STRESSES

Even though the design factors built into a building are uppermost in the mind of the architect who is creating the structure, one factor that you as a firefighter must always consider is the degree of structural abuse that occurs after the building is built and in daily use. There are several structural abuse factors that can degrade the work of even the finest architect:

- Subjecting the structure to loads for which it was not designed

- Structural modifications by unqualified workers or contractors

- Deterioration

- The forces associated with the violence of a fire

It is important for you to ensure that the materials specified on the plans are the materials actually used in the structure. There have been numerous instances where the lack of a critical structural element, building material, or member has led to the death of fire suppression personnel.

You cannot know the effects of many of these criteria if your fire department fails to conduct inspections and pre-incident planning visits. If you do not get out and visit these buildings, how will you know what modifications have been made, particularly if those people doing the work failed to secure the proper building permits? You can anticipate both the fact that a building will deteriorate as it gets older and the reality that any fire will unleash the devastating forces of combustion as listed above. Never fail to consider these things as you conduct your sizeup of a fire scenario.

Let us take a look at physical stresses as they affect building construction. It is essential for the practicing fire officer to know how buildings are put together and what types of stresses they will endure. To that end, we will speak to the way in which buildings are built. As we stated earlier, IFSTA emphasizes in its **Building Construction** manual that, "a building is viewed as a system which provides an environment to enhance or support human activity." To that end, we as fire officers must study any attack upon such a building system in the same manner in which a physician diagnoses a disease that is attacking the human body. It is just that critical.

Problems Caused by Construction

The first three things to consider in any structure are size, type of construction materials used, and methods used in building it. The configuration that the architect chooses for these factors is also important. Consideration of these factors should be a part of plans review and inspection programs. They also serve the pre-incident planning and standard operating procedures programs. An excellent example of how construction features can affect fire operations lies in the current trend toward total environmental control in buildings. Since all heat and air conditioning are internally controlled, fewer windows are provided. If a problem develops with the installed emergency ventilation systems, firefighters are in trouble.

Another problem found in construction comes from the alteration of architectural plans by contractors. An example might come from substituting a lower, more flammable grade of sheet rock. This can allow for a more rapid deterioration of structural integrity, and a faster fire spread. Substituting a combustible building material for one that is supposed to be noncombustible can also cause an unanticipated structural failure. These are just a few of the many problems that can occur. These mistakes have to be caught by inspection personnel in order to prevent future problems (Figure 18.14).

Serious fire problems are also called by poke-throughs. Fire can move through a hole created by a plumber, electrician, or a cable television installer. These are the type of errors that wreak havoc with the building construction system envisioned by the architect.

Problems Caused by Alterations

Alterations made to a structure can have a profound effect on its stability and fire resistance. Alterations frequently make structures more prone to collapse and cause firefighters serious difficulties in search and rescue operations. Let us take a look at some of the alterations that building owners and tenants might undertake that can have fatal consequences:

- Void spaces that can conceal the spread of fire or smoke
- Remodeled ceilings (where the old one is not removed)
- High-density fiberboard
- Flammable adhesives
- Highly flammable, non-rated wall and floor coverings
- Removal of supporting columns
- Added dead loads
- The use of water pipes for columns and beams
- Substandard building materials

These are just a few of the many things for which you must constantly search in your pre-incident program. If this seems like a great deal of work, you have been paying attention.

Deterioration

Deterioration of a structure can occur as a result of lack of maintenance or the natural forces of nature (Figure 18.15). A series of bad winters, with repeated heavy snow loads, can cause roof supports to weaken. People often make a practice of taking bricks out of older buildings to sell them or use them elsewhere (Figure 18.16).

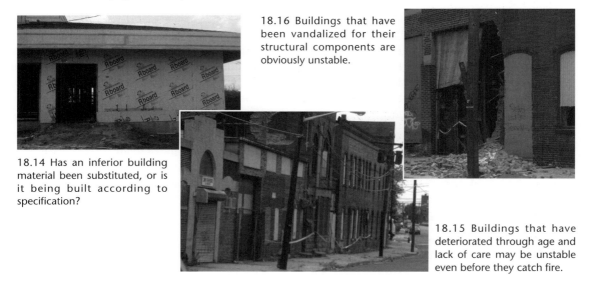

18.16 Buildings that have been vandalized for their structural components are obviously unstable.

18.14 Has an inferior building material been substituted, or is it being built according to specification?

18.15 Buildings that have deteriorated through age and lack of care may be unstable even before they catch fire.

Overloading

Overloading of a structure can occur in a number of circumstances. This can happen in an accidental way, over time. Successive occupants often add their personal touch to the building, and eventually the weight is too great and collapse occurs. Overloading can also come as a direct result of a conscious violation of engineering principles. Careful plans review and inspection will be needed to find and correct these errors.

Stresses Due to Fire Fighting Operations

While you, as the fire officer, have no control over the application of forces external to your fire fighting force, you must always keep in mind that these forces are at work. You must caution your firefighters to do nothing that will make matters worse. An example of this would be overloading an obviously damaged staircase or roof.

Backdraft

There is one additional point to ponder, and that is the effect a backdraft can have upon an otherwise sound structure. If you think of a backdraft as being an explosion, nothing more or nothing less, you will understand why it is so important to stop it from occurring. Regardless of how well any building is put together, if you have an explosion, you will have tremendous damage. The type of construction, materials used, and configuration of the building can make the effects of such an explosion greater or lesser. Needless to say, a backdraft is a tremendous stressor that can have catastrophic effects, including the destruction of the building and all of its integral systems.

COLLAPSE

How many times have you picked up a professional fire journal and stared at the glaring headlines that trumpeted the death of another firefighter in a building collapse? We want to do something to halt this senseless loss of life, as well as the resulting flurry of injuries. It is then critical to become aware of the clues a building can give to you in terms of its state of physical health. And once you have learned all of these things, you must then **REMEMBER THEM** and **APPLY THEM** as necessary at emergency incident scenes.

One of the greatest learning tools is personal experience, such as surviving and profiting from building collapse. I know a bit about this topic. I have been the victim of a floor collapse in a residential house fire, and I fell through a missing floor many years ago. These and other events have given me a heightened sensitivity regarding the matter of structural integrity.

Do not be lulled into a false sense of security based on a building's fire resistance, either in terms of construction features or automatic fire protection systems. The seven areas listed below are the prime candidates for collapse; focus your training efforts on them. Problems in these areas can create fatal defects that can cause a structural collapse. During your pre-incident planning sessions, you must note the pertinent data on each of these structural areas.

- Exterior bearing walls
- Interior bearing walls
- Columns
- Beams, girders, trusses, and arches
- Floor construction
- Roof construction
- Exterior non-bearing walls

While it is nice to have automatic fire sprinkler protection in place, be aware that the protective factor has never been 100 percent. Remember that, given enough heat, just about everything will burn, decompose, or crumble under the attack of fire. Further, slip-ups can happen during any of the phases of planning, design, and construction. No matter what the plans say, mistakes can happen. There are also a number of problems with modern, lightweight truss construction. They burn quickly and fail. My department's first contact with this problem came at a fast-food restaurant fire back in 1981. The destruction that occurred in the areas where the metal gusset plates were attached to the wood was tremendous.

Clues to Look for Regarding Collapse

It would take a whole book to give you every clue in terms of building collapse. Here are a dozen to get your interest up:

- Missing or damaged bricks in walls (Figure 18.17)

- Sand lime mortar in old and restored buildings (This can be discovered by checking to see how well the mortar happens to be clinging to the bricks. If the brick is clean with very only minimal mortar on it, there is a high probability that the older, weaker mix was used.)

- Wooden lintels must be suspect and can usually be found only during pre-incident planning visits.

- Cracks in walls (Figure 18.18)

- Star attachments on outside walls indicate that the building is held together with tie rods that can fail during exposure to fire.

- Any arch is a potential candidate for failure.

- Hollow clay tile has a higher potential for failure and collapse.

- Columns built into the wall indicate basic instability.

- Columns built inside of existing walls

18.17 Buildings that are missing parts of their walls or roofs are likely candidates for collapse.

18.18 Walls that have separated are a good clue that the building is very unstable.

- Holes in walls without supporting lintels

- Hanging signs and theater marquees can fail easily, with disastrous results.

- Cantilevered balconies are candidates for fire-related collapse.

There are three operational areas in your fire department that must work closely together in any collapse prevention program:

Fire Prevention Division: prevent it from being badly built; correct any improperly planned alterations; enforce the fire code

Training Division: teach the skills and knowledge necessary to arm your operational forces with an awareness of the problem

Suppression Division: learn the necessary skills; use them in actually visiting and pre-incident planning area buildings; know it all when you are called to fight fires in your district

I suggest that you move forward with this study of building collapse immediately. You can never tell when that NEXT TIME will occur.

SUMMARY

In the world of fire fighting, it is what you do not know that can kill you. This is particularly true when it comes to the subject of building construction and structural support mechanisms. When the structural support members in a building fail, the resulting collapse is unforgiving. Learn what holds a building up and how to notice that one is coming down.

A thorough knowledge of basic building types is a necessity if you are to provide for the safety of your firefighters. You must learn about the basic building types. What are the basic features of any structure? Look at such things as construction materials used, configuration of the layout, total building size, and methods of construction. Each has a role to play in any fire fighting operation.

We have also discussed those factors that can cause a building to fall down around your ears. There are the stresses of heat and explosions. And do not forget the curse of faulty construction, age, lack of maintenance, and repeated alterations. Remember also that fire fighting operations themselves can be tremendous stressors.

Translate your book knowledge into community-oriented experience through the mechanism of community visits. Learn what might cause a structural collapse in your town, and use the knowledge to develop General Operating Guidelines that help ensure the safety of your firefighters.

When you ask the question "What have I got?" during your sizeup operations, you should also think construction:

• Weather • Incident type • Time • Height

(WITH) + Construction

19

Firefighter Safety

People are our most important product. Their safety is an issue critical to the future success of your fire department. But safety can only occur if the light of a brilliant idea comes on in someone's head. Preferably that light bulb should come on in the head of someone at the highest organizational level, someone who has the wherewithal to get an important task accomplished. Unfortunately, the people at the top are frequently insulated from the problems of the masses.

Either by the intent of those being led, or personal desire, fire service leaders sometimes grow out of touch with the people in the fire stations. This is a dangerous trend. However, it is one that I have observed on far too many occasions. It is also my opinion that some people have not shown sufficient concern for the troops. They have done this by making the safety officer's position a hollow shell. It has been used, in some departments, as a place to park people after their promotion to a higher rank, marking time until a field command position became available.

This type of nonsense does NOT meet the intent of the standard and can serve as the basis for a tremendous law suit should someone be injured by the improper actions of an unqualified safety officer. Do it right!

(**NOTE:** For more information on safety, I strongly recommend that you obtain the IFSTA text on **Firefighter Occupational Safety**. It will cover each area mentioned in this chapter in much greater detail. I urge you to make safety a part of your life. All that has been taught in this text revolves around the safe application of the information. The alternative is not an appealing prospect.)

However you decide to address safety issues, please keep one thing in mind. Everything you do should support the development of a safety mindset in your people. The best safety program in the world will not be effective if your people do not have the knowledge or willingness to get the job done safely. It is your job to push safety.

In order instill this safety mindset in your people, your safety program must be comprehensive; and most important, supported from the upper levels of the department. Realize that the person ultimately responsible for safety is YOU. Be responsible to yourself and those under your care.

DEVELOPING A SAFETY STRUCTURE

Having made the first conscious decision to pursue the illusive ghost of organizational safety, the journey should next move into the world of the code book. There are a number of important tasks that must be done by the fire department in order to develop a safety structure for the organization:

1. Develop a Fire Department Organizational Statement

2. Create a risk management plan

3. You will have to create policy (an organizational way of doing things.)

4. Define roles and responsibilities

5. Appoint a department safety officer

6. Set up a committee

7. Develop a training program

8. Develop a vehicular training program

9. Buy safe, train safe, keep it safe

10. Acquire appropriate protective clothing

11. Develop rules for emergency operations

12. Ensure facility safety

13. Provide a medical and physical Program

14. Provide member assistance

Guidance in establishing a safety program comes from NFPA 1500, *Standard for Fire Department Occupational Safety and Health Programs*. This standard was designed with a specific purpose in mind. As stated in section 2-4 of NFPA 1500, " ... (T)he fire chief shall appoint a designated fire department safety officer. This position ... shall be responsible for the management of the occupational safety and health program ... (and have) such additional ..." people and resources as the department deems necessary to do the job well.

One of the most important concepts identified in the administrative section of the code deals with the concept of equivalency. In drafting this document, the committee determined that there were a number of ways in which people could be trained to fulfill the requirements of being a safety officer. The key to success is to have competent people performing the safety function, regardless of the manner in which their competencies are gained. The ground rules that you must set up in order give legitimacy to your department safety officer come from paragraph 2-1.3 of NFPA 1521, *Standard for Fire Department Safety Officer.*

The onus for the total safety program is placed directly where it belongs: on the back of the local Fire Chief. It states therein that, "(T)he Fire Chief shall have the ultimate responsibility for the Fire Department Occupational Safety and Health Program as specified in NFPA 1500, *Standard for Fire Department Occupational Safety and Health Program.*" This is just as it should be, because the chief executive of any organization must accept the culpability, as well as the credit, for whatever happens in the organization.

In far too many instances, fire chiefs attempt to duck out on the unpleasant chores in order to make more time for the good things. If your local safety officer program is to have any chance at success, it must be fully, and forcefully, supported by the Fire Chief. And this support must be in written form.

Appoint a Department Safety Officer

NFPA 1521 tells us that the candidate for a safety officer's position must be an OFFICER. This requirement may seem like a fairly basic starting point, but it has a meaning. The people who wrote the standard felt that a safety officer could not be a firefighter. The designated safety officer position responsibilities are great (Figure 19.1). The demands for interface with fire department personnel at every level are frequent. Cease-and-desist orders must be given under certain tightly defined circumstances. Because of these factors it was felt that the position required an officer.

The person charged with safety must know the boundaries of their world and become proficient at all tasks within their field. According to NFPA 1521, *Standard for Fire Department Safety Officer*, the "(S)afety Officer shall have the responsibility to identify and cause correction of safety and health hazards." To most people, the fire department safety officer's role is limited, in their minds at least, to a fireground scenario.

They envision the safety person as some sort of a free-lance White Knight. They see a person who sallies forth to battle the unsafe acts found during our pitched battles with the RED DRAGON of fire. The safety officer is imagined to be sort of a Mother Teresa for the poor unwashed masses out there fighting uncontrolled fire. This is all true, as far as it goes.

Safety officers have an extremely responsible position within their fire departments. People will continually look to them for guidance in the "right" way of doing things. While many officers limit their role to the fireground, there are many more areas of concern.

Emphasize Training

Training is one area that does not receive enough credit for being a part of a fire department safety program. If people are taught the correct way to perform their duties, they will operate in a safer manner, with a lesser chance of injury (19.2). In-service drills and periodic refresher training keep people going in a positive direction. If you teach people proper techniques and monitor performance, your efforts can limit injuries and possibly prevent death. Ensure that whenever any sort of training is occurring, it is performed in a safe manner. Here are just a few examples to stimulate your creative juices:

- During a company-level driver training program, are the participants wearing hearing protection?

- Have your pumpers been service tested to ensure that they should not fail during live fire training exercises?

- Are the classroom areas in your fire stations properly vented?

- Have your smoke house and burn building been checked by a licensed professional mechanical engineer lately?

- Are your personnel wearing their personal protective equipment during drill sessions?

19.1 The safety officer must have the ability and authority to evaluate operations on the fireground. *Courtesy of Ron Jeffers.*

19.2 Firefighters will not operate safely on the fireground if they have not been trained to do so. *Courtesy of Ron Jeffers.*

These are just a few of those instances where your role as a safety officer can have an impact far beyond today. Each of the items listed above, and the countless ones that I hope you will identify, can add years to the lives of your personnel.

Keep Records

Just being safe is not enough; you must be able to prove it. Section 3-1.1 of NFPA 1521 specifically states that, "(T)he fire department shall maintain records of all accidents, occupational deaths, injuries, illnesses, and exposures in accordance with Section 2-6 of NFPA 1500, *Standard on Fire Department Occupational Safety and Health Program.*"

This is a fairly straightforward statement. You must keep records that document everything related to occupational safety and health (19.3). The standard also goes on to pre-scribe the Safety Officer as the individual charged with this task. In the same paragraph, the standard specifies that the Safety Officer shall manage the collection and analysis of this data. Just how will you accomplish this critical task?

19.3 The safety officer is in charge of keeping records of training.

As a basic beginning, you must establish an organized system of reports and files. If yours is a small department, with minimal runs and a fairly fixed level of personnel, you can get by with a pen and paper, or typewriter system. However, even the smallest of departments can benefit from the use of a computerized database.

Participation in Committees

The safety officer also holds a critical role within the fire department hierarchy. Nothing impacting any aspect of safety should occur without the knowledge and guidance of the safety officer. This issue is covered under Section 3-2 of NFPA 1521, which specifies the components of the safety officer's liaison role.

Transmit Information

A key fundamental of this role is the safety officer's membership on the departmental safety and health committee. Whether the safety officer serves as a member or the chair does not matter, but he or she must be a strong participant in all safety and health-related issues.

The safety officer is also charged with the reporting requirements of the committee. This individual must also ensure that the Fire Chief or his designated representative is aware of what the safety committee is doing.

Safety is not an area where surprises are appropriate or desirable. These reports are generally made on either a monthly or emergency basis. The monthly report should cover all actions of the safety committee and the safety officer in a given month. This can be forwarded along with the necessary reports on deaths, injuries, and illnesses. The safety officer must take great pains to ensure that his work does not proceed in a vacuum. Upper level support is critical to the success of any safety initiative.

Another critical component of the liaison role involves forwarding safety suggestions and recommendations to the Fire Chief. These things will need his approval for implementation, so get them to him as quickly as possible.

It is up to the safety officer to provide assistance to officers and firefighters in their district survey programs. You must do this so that they will be able to identify and report safety and health hazards in their respective districts. These hazards may well have an adverse impact upon operations, so they need to be addressed.

Safe Equipment and Apparatus

The safety officer should also serve as the conduit for information to flow from line personnel to the staff officers charged with new apparatus and equipment acquisitions. This is critical if operational hazards are to be eliminated. Development of a vehicular training program is critical to the future of safe operations in your department. Training in defensive driving and safe vehicle operation can pay benefits in reduced accident levels, as well as lowered vehicular injury rates.

Buy your people safe equipment, then train them in safe operating techniques (19.4). This will help reduce injuries, increase fireground efficiency, and allow you to have a better chance to keep department operations safe for the members. Protective clothing should be acquired that meets the appropriate state and national standards. Your people should be trained to use it properly. And it should be maintained in usable condition.

You have to concentrate on enforcing the role of safety within your training program. The greater your emphasis on safety, the better the chance that your emergency operations will be conducted in a safe way.

19.4 Make sure that apparatus is as safe as can be. *Courtesy of Ron Jeffers.*

Facility Safety

Facility safety is an issue that is frequently overlooked. Poor ventilation, lead paint, rust in the water pipes, and poor environmental conditions can injure your people. But not in a sudden and devastating way. It will occur in an insidious, secretive manner. You must pay attention to where your people live and work.

Medical and Physical Programs

Medical examinations for all personnel and a departmental physical fitness program can go a long way in heading off line-of-duty deaths and injuries. If you hire people who are healthy, and work to keep them that way, you will get better service from them. And the overall costs for medical care will be lower.

Another way to make sure that your people are ready for duty at all times involves the provision of an employee assistance program. Smoking, over-eating, alcohol and drug-related problems will take a toll on your service delivery capability. If you help your people, you may not lose them. This makes sense.

Be sure that all of your operating personnel receive a medical examination on a yearly basis. It is best if you begin with a thorough entrance physical to establish a baseline for assessing an individual's physical condition. In this way each annual physical charts the effect of fire fighting on the individual. And it can provide appropriate warning if an individual begins to develop job-related physical problems.

Having recruited healthy candidates into your department, it is up to you to see that they stay that way. One good method is through the use of a physical conditioning program. Counseling on diet and lifestyle are also important. Bad habits, whether on or off duty, can have a negative impact on job performance.

SAFETY ON THE FIREGROUND

Organizational efforts are one thing. However, it is on the fireground where unsafe acts have a great potential for maiming or killing our troops. What are some things that you must do at every fire to keep your people safe? Because most injuries occur at the scene of fires, we will concentrate on this area:

- Fires are not totally unpredictable; however, if firefighters and officers ignore what a fire might do, they stand a good chance of being injured.

- If it is cold, dress warmly. Watch where you are walking and be careful on ice. We shoot lots of water and nature ensures that it will freeze in cold weather.

- If it is hot, conserve your efforts to the greatest extent possible. Rotate your operational forces frequently. Ensure that fluids are available. Be sure that your local EMS component is available to monitor the physical condition of your people.

- Use self-contained breathing apparatus. And do not limit your use to interior operations. Vehicle fires can generate dangerous levels of toxic smoke.

- Monitor the fatigue level of your personnel. Tired people are more apt to become injury statistics.

- When using forcible entry tools, keep an eye out for people who feel they must wander near to you (Figure 19.5). They may have an unconscious desire to be struck on the head. Do not help them along.

- When climbing ladders, be sure that you have someone securing the base of the ladder as you climb.

- Do not overload ladders. Refer to the manufacturer's specifications for guidance.

- Be aware of the fire building. Does it have a bowstring truss supporting the roof? Might it have one of those new rubber roofs that can hide a multitude of structural sins? Or could it be a lightweight truss or metal pan deck, just waiting to drop an unsuspecting firefighter into the blaze?

- Do not daydream when it comes to assessing construction. Are there cracks in the walls? Are the floors sagging? If you are up on the roof is it solid or spongy? Or is it a new roof style where you cannot tell the difference?

- What is the potential for collapse? There is no sure guide as to whether a wall will fall a little or a lot. We always treat them like it will be a lot, if it happens. It's safer that way.

- Keep your aerial devices away from electric lines. Be at least 10 feet away, more if they are high tension. Not only does it make good sense, it is the law.

- Keep your breathing apparatus on during overhaul.

19.5 Operations involving forcing or striking tools are inherently dangerous. Make sure that everyone knows how to handle tools as safely as possible. *Courtesy of Ron Jeffers.*

- Give your teammates space when they are using hooks and axes to overhaul a fire building. They need room to maneuver.

- Pay attention to what you are doing. Daydreaming can be dangerous.

The road to safety is a long and winding path. The journey must be made by all parties concerned. It cannot just be a paper plan of the leaders, or an informal plan created by the people fighting the fires. There is a great deal of hard work and heartburn in working together to develop an official safety program. But it beats the hell out of hurting or killing people.

20

Sprinklers and Standpipes

We in the fire service tend to think of ourselves as the weapon of choice when it comes to combating the dangers of fire. Tough firefighters advancing in close combat against our common enemy: FIRE. Unfortunately, this is a poor view of reality. We must take the high road if we are to truly become the fire service of the future.

Fire departments that are convinced that manual fire suppression is the only game in town will never provide the best service for their community. We must become advocates for the best weapon in the fire protection arsenal: automatic fire sprinkler systems. Sprinklers are like little firefighters waiting on their branch lines to attack fire as soon as they feel it. How much more efficient can any system of protection be? There is an instantaneous response from above, applied in a measured spray, according to a design criteria.

AUTOMATIC SPRINKLERS

Automatic fire sprinkler systems are one of the most dependable means at hand for controlling fires. Early efforts at developing a means of applying water directly upon a fire during its earliest stages revolves around the use of perforated piping. These systems came into use in during the period 1850-1880, with their most frequent use group being the various mill facilities used by the industries of that day. The next step beyond the perforated pipe involved the use of open sprinkler heads. One problem in particular with both the perforated pipe and the open head involved the problem of getting the water started when fire occurred.

It was not until 1878 that we find the first pragmatic employment of this practical methodology. This was the year when the first Parmelee sprinkler was installed. Although it bears little actual resemblance to our modern systems, it was quite an improvement over existing technology. Just what is this thing that we call an automatic sprinkler system? Like any other type of system, it is the sum of a number of important parts. Let us first look at the fact that there are a number of different system types:

- Wet-Pipe
- Dry-Pipe
- Special Arrangement

As the name implies, a wet-pipe system has water ready and available in the pipes for immediate use when a sprinkler is fused, or opened (Figure 20.1). A dry-pipe system has water available at the dry-pipe valve, ready to be released into the system when a sprinkler opens.

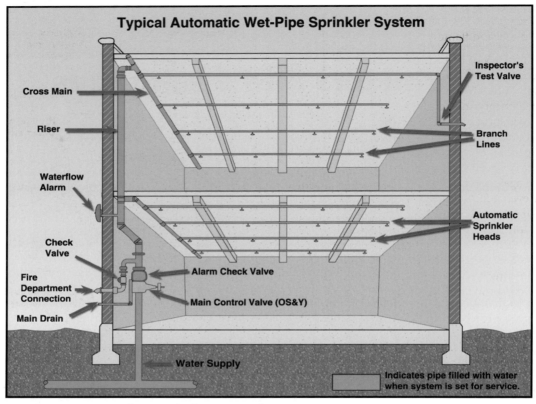

Typical Automatic Wet-Pipe Sprinkler System

Cross Main

Riser

Waterflow Alarm

Check Valve

Fire Department Connection

Main Drain

Alarm Check Valve

Main Control Valve (OS&Y)

Water Supply

Inspector's Test Valve

Branch Lines

Automatic Sprinkler Heads

Indicates pipe filled with water when system is set for service.

20.1 The wet-pipe system has water in the piping at all times.

When a sprinkler opens in a dry-pipe system, the system's air is released and the water is allowed to flow past the control valve out into the system for release through the sprinklers (Figure 20.2). Devices known as exhausters and accelerators are employed to assist in quickly moving air from the system. The accelerator quickens the operation of the dry-pipe valve and, as its name suggests, the exhauster works to exhaust system air to the outside environment.

For certain specialized situations, pre-action systems are employed. These are essentially dry-pipe systems equipped with heat-actuating devices that sense the presence of heat and allow the water to flow into the system for distribution from the sprinklers when they are fused. These systems are generally used where cold is a problem, but greater speed of water discharge is needed to control or extinguish an incipient fire.

It is important for you to understand that a basic assumption of sprinkler protection rests upon the theory of the automatic discharge of water. This water must be supplied in sufficient pressure and density to control or extinguish a fire in its earliest stages. In many cases, a system can hold a fire in check pending your arrival. This can be of great assistance. Better to have a small fire, getting wetter by the minute, greet you than a roaring inferno at the base of a heavy column of smoke and flames.

A review of historic fire protection records indicates that fires are generally controlled or extinguished by a small number of sprinklers. This same review speaks to the fact that their success can be established in well over 90 percent of the cases where systems have been called upon to protect property against fire. Life safety success is an even more important measure of their success than simple property protection. There is no record of a multiple death fire (a fire that kills three or more people) in a completely sprinklered building in which the system was operating properly. The exceptions to this come from explosions or flash fires. Not a bad record when you consider the number of fires that happen each day.

It is important for you to understand the various reasons you can use to sell members of your business and residential communities on the concept of installing or retrofitting sprinkler protection in your community. They fall into the following broadly defined areas:

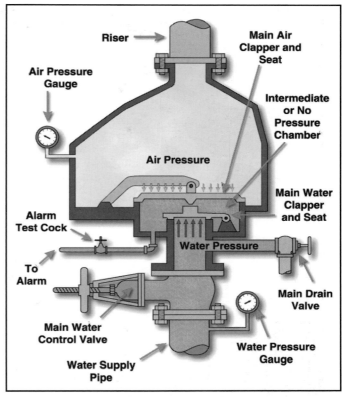

20.2 A dry-pipe system allows water to flow into the system when sprinklers activate.

- Increase life safety

- Minimize business interruption

- Limit loss of property

- Reduce insurance premiums

- Proven success rate (Only 1 in 16,000,000 installed sprinkler heads is estimated to fail)

- All sprinklers DO NOT open when the system discharges

    NFPA standards with which you should be familiar are as follows:

- NFPA 13, *Installation of Sprinkler Systems*

- NFPA 13D, *Installation of Sprinkler Systems in One and Two Family Dwellings*

Dwellings

- NFPA 13E, *Guide for Fire Department Operations in properties protected by Sprinkler and Standpipe Systems*

- NFPA 13R, *Installation of Sprinkler Systems in Residential Occupancies up to Four Stories*

- NFPA 14, *Standard for the Installation of Standpipe and Hose Systems*

It is up to YOU to learn as much as you can about automatic fire sprinkler systems. In order for these systems to be fully effective, you must support them in times of emergency.

A primary rule to remember during fire fighting operations is to feed the system. Do this according to your department's operating guidelines. Many times a sprinkler system can hold a fire in check with its own water. However, to provide the fullest capability, you must supply additional water under pressure (Figure 20.3).

STANDPIPES

It is now time to take a look at the installed standpipe. At its most basic, a standpipe system involves vertical piping with outlets for fire fighting hose. These systems can be either filled with water (wet-pipe) or filled with air (dry-pipe). Each of these systems has but one aim: to provide water for manual fire fighting. But, you ask, don't we have fire hoses attached to our pumping units for that purpose? True enough for the average residential fire in a one-, two-, or three-story building. However, when the buildings are tall or complex, we need a bit of help. Where they are installed, standpipes can make our job faster and easier (Figure 20.4). It is important to ensure that they are properly installed and maintained.

20.3 Firefighters supporting an outdoor standpipe. *Courtesy of Mike Wieder.*

20.4 Operations can be helped by hooking into the standpipes.

Standpipes originally came about as a practical response to the problems brought about by the invention of the elevator. When buildings began to grow taller, the problems began. After a few fairly destructive fires, it was noted that firefighters were having a difficult time getting water to the upper floors of these new creations of construction technology. Someone proposed to make things easier by running a pipe up the side of a building. Firefighters could then hook up a feed hoseline to the base and pump water up to the fire floor, where another outlet allowed them to hook up their attack hoseline. However, a problem was soon noted: These outside pipes were subject to the whims of the weather. When it was warm, things worked relatively well, but cold weather created an entirely different ball game. The pipes froze and burst, rendering them useless. Further, designers frequently overlooked or miscalculated the pressures that were needed to raise water to an elevated area within a building. Pipes burst under the strain and firefighters had no choice but to carry hose up in the building (if they could). This apparently was still a problem when the Home Life Building fire occurred in New York City in December of 1898. A system failure occurred when pressures were applied to the standpipes. The pipes burst, indicating that proper attention had not been given to the pipe specifications.

Over time, these and other problems were corrected. The basic principles of standpipe use have been brought into the National Fire Protection Association's National Fire Code. Their design, installation, and use are governed by NFPA 14, *Installation of Standpipe and Hose Systems*. According to the latest edition of this code, standpipe systems come in three distinct types:

Class I Systems — provide 2½-inch hose outlets for use by fire department personnel

Class II Systems — provide 1½-inch hose outlets for first aid fire fighting

Class III Systems — provide features of both systems by using 2½-inch outlets with pressure reducers

Regardless of the class of system found in a building, the basic requirement exists for water to get the job done. The demand for a standpipe system is a function of several variables:

- Required flow rates and pressures
- Required flow rates for additional standpipes
- The water supply requirements for automatic sprinklers sharing common piping
- Required flow duration

Many standpipe systems have a pressure-reducing mechanism located at each standpipe outlet in the building (Figure 20.5). This allows non-fire department personnel, who are not fully familiar with hoseline operations, to use the first aid fire hose. You must know where these are located and how to remove or disable them. A failure to do this will limit your ability to successfully use this important, built-in fire protection system.

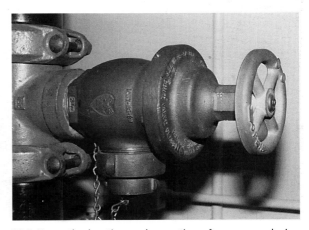

20.5 Know the location and operation of pressure-reducing devices.

As with just about every other aspect of fire fighting operations, the best time to become familiar with any standpipe system is during your pre-incident planning visits. You must also be able to find the system siamese pump-in points, as well as the fire hydrant nearest to each one. This will save time and steps during a fireground operation. If possible, you should schedule drills using the system. Be innovative and be creative. Don't let the next fire in a protected building be the first time you use such a system. Be proactive and be ready.

SUMMARY

The use of installed fire suppression systems can materially affect the outcome of your fire fighting operations. Learn how these systems work. Drill in supporting them and know where they are located. They can save the lives of both civilians and firefighters. They can answer a great question within our system:

What have I got to stop it?

The answer: Automatic Sprinklers and Standpipes

21
Decisionmaking & Problem Solving

One of the critical elements in your preparation to be an effective Incident Commander comes from the world of management. You must be able to make weigh facts, evaluate circumstances, and make an effective decision. In this chapter we will provide the means for you to learn about problem-solving and decision-making.

Many people assume that making a decision is a very simple process. It is and it isn't. Understanding the process is your first task. You will then have to practice the process. Then you will face situations where you are compelled to recognize the need for a decision. Basically, a decision is the choice of an action in response to a problem. We define *problem*, in this case, as a deficiency in our needs, which occurs in a given situation under a specified set of circumstances. To be classified as a decision, you must be required to make a choice. If there is no choice, there can be, by definition, no decision. The unfortunate problem that many people face is that the lack of a choice causes them to want to find a choice. This creates a problem, and then these people are happy. In order to assist you with improving your decision- making capabilities, we will examine the process of making a decision. We will look at each of the steps and then help you to understand their use.

TYPES OF DECISIONS

There are two types of decisions that you will be called upon to handle. The first is the routine decision. These are normally simple matters. They tend to be periodic and repetitious, and have fairly certain outcomes. Many common fireground decisions fall into the routine category. There should be no question as to whether we will use the or Incident Management System. This is usually perceived as the best way to operate. More than three decades of research exist to verify this fact. And in many jurisdictions, its use is mandated by government edict. Other decisions that are normally pre-made for us include such things as:

- Which units respond to which types of incidents?

- When is a rescue company dispatched to a working fire?

- How many pumpers and how many aerial ladders are due to respond to a target hazard in your community?

This is just a short list. Undoubtedly you can come up with a great many more. The object here is to make things as easy as possible for your troops by arming them with pre-made decisions, or general operating guidelines. Just add fire and water and you have a decision.

BRAINSTORMING

An excellent way to create these guidelines and pre-made decisions comes through the use of a process known as brainstorming. This is a freewheeling process used to generate as many solutions to a problem as possible. But it is done in such a way as to discuss ideas for these solutions without imposing judgments on people. You do not want to turn them off with criticism. You and your people are encouraged to present your thoughts as you get them. You are discouraged from making fun of anyone's ideas. All ideas are written down and used in the process. Many times ideas build one upon the other. What one person perceives as a silly suggestion may be turned into the best idea possible through the group interaction process.

Remember that the keys to effective brainstorming are simple. But ignore them at your peril. They are:

- No criticism
- Freewheeling discussion
- Quantity of ideas is desired
- Combination of ideas and improvements are sought
- Write things down so that ideas are not lost

Once you become good at brainstorming, you can apply these skills to the creation of a set of general operating guidelines (GOGs). These will be available to handle your most commonly encountered fireground problems. In this way, your personnel can be prepared to respond to future problems by having guidelines available to them beforehand (Figure 21.1).

So much for the routine decisions. There will be those cases when you have to unleash the full artillery of the decision-making process. There will be those occasions when you face a number of alternative choices for a non-routine decision. The incident may involve a large and complex structure. It could be a type of transportation incident that you have only read about. Or it could be the sudden onset of a structural collapse. Each of these is an event that you do not see every day. In this case, you would be well-served to use the following steps:

1. Define the actual problem.

Be sure that what you are looking at is the actual problem and not a symptom of the problem. Be sure that you are treating the pinched nerve that is causing the headache and not just taking aspirin for the symptoms.

2. Collect information that will assist you with developing solutions.

During a fire, such information as wind speed and direction, smoke color, and flame conditions can give you clues to the problems you may face. In this way, you can prepare for solving problems that have these matters as elements.

3. Generate alternative options.

As you look at a large fire, on a windy day, you begin to use our fireground sizeup method (Figure 21.2). The "What have I got?" question leads you to the following alternatives:

- Should we use an aggressive interior attack?
- Should we keep to the block of origin?

21.1 The department that makes and follows GOG's will operate more efficiently in all types of situations. *Courtesy of Ron Jeffers.*

- Should we use a blitz attack and move in?
- Should we use a blitz attack and think about moving in? (Figure 21.3)

21.2 Use all the information you can generate to formulate your plan of attack. *Courtesy of Harvey Eisner.*

4. Evaluate the alternative options.

By looking at the fire and pondering your options, one will become more practical as time passes. Unfortunately, at a fire you cannot let a lot of time pass, because of the destruction that is ongoing during the decision process. Many people get hung up at this phase. They are so busy pondering all of the possibilities that they never make a decision (Figure 21.4).

5. Pick one. Get off the dime and select an alternative.

6. Do it.

7. Check your feedback to see how the decision is working.

The object here is to evaluate the success of your choice. If it is getting the job done for you, then your decision is correct. If it does not solve your problem, choose another alternative and see how things go. If none of the alternatives work, you may have to start the process over from the beginning (Figure 21.5).

I can remember a fire where the problems became so overwhelming that we pulled everyone out of the block where the factory was burning. It was so unsafe to continue what we were doing that we regrouped our forces two blocks away and went back to a totally different method of attacking the fire. Eventually, we did manage to extinguish it. And more important, none of the more than 80 people fighting the fire were injured. That spells success to my way of thinking.

REASONS FOR BAD DECISIONS

Regardless of our intentions, bad decisions occur from time to time. Why is this the case? Many times people charged with making decisions do not trust the people who work with them. It also

21.3 If you need big water, use big water – and quickly.

21.4 Once you have weighed plans and risks, do not hesitate to carry them out. *Courtesy of Harvey Eisner.*

21.5 If you are facing something really big, do not hesitate to attack defensively and call for more help. *Courtesy of Harvey Eisner.*

could be that they have not trained their subordinates to make decisions or ever allowed them to develop as leaders. In any case, they ignore what people have to say and make all decisions themselves. They ignore data that does not confirm their way of thinking. Bad decisions can result from this myopic view of life. Rather than soliciting advice and opinions, these people rely on their own preferences. They presume that they have seen and experienced everything. Quite simply, they say the heck with everyone else. I know what is best and the rest of you do not. This type of thinking can lead to serious consequences.

Sometimes people make decisions based strictly upon tradition. These are the people who live by the motto, "We've always done it that way." These are the people who will always seek to use a 1¾-inch hoseline for every type of fire, from a trash can to warehouse. This is not a good way to operate.

Worst of all are the people have not learned from the bad decisions they have made in the past. It is their feeling that if they keep trying the same decision over and over, it will eventually be the right one. People can be killed in this fashion. Some people truly never do learn from their mistakes.

SUMMARY

As an Incident Commander you are required to make decisions of all kinds. You will succeed if you learn the following procedures:

1. Define the problem.
2. Collect information.
3. Generate alternative options.
4. Evaluate the alternatives.
5. Pick one.
6. DO IT!!!
7. Check your feedback to see how the decision is working.

Absorb the meaning of what it takes to make a decision. Practice making decisions. You will get better with practice.

22

Some Philosophical Observations for My Friend — The Reader

This text was written with a single purpose in mind: to share my decades of fire fighting experience with other people who are engaged in the fire fighting business. I have had the good fortune to have attended literally thousands of fires and emergencies of all types and kinds. I have made a lot of mistakes, which is what a human being does. What I wanted to do is give you the results of my education at the University of Hard Knocks. You may recognize some of what I say, and feel that you have read it before. There is a reason for that. A person's collective experience is a combination of the following:

- The drill ground

- The textbook

- The fireground

What I now share as a philosophy with you is a distillation of three decades of working that process. I never met the late Lloyd Layman of West Virginia, but have read several of his books. I often discussed fire fighting with the late Bill Clark; he was a friend. I have discussed many facets of the fire service with Alan Brunacini; he is a mentor. And there are countless others who have shaped me to be the fire officer I am today. Hopefully, I can have some positive influence on your future endeavors.

This book has been written with no single audience in mind. It is not a book solely for career or volunteer firefighters, or for industrial or military firefighters, for that matter. It is for people who go to fires, get dirty, and create extinguished fires. PERIOD! This book is targeted to the operational fire officer. I also know that the average firefighter likes to know what's going on. This book is for that whole range of hands-on people because they are the ones who are called upon to handle a whole range of problems from year to year. For everyone, I tried to keep the language and the concepts simple. It was my intention to create a decision-making routine inside your brain. Each time you arrive on scene, I want you thinking in a conscious, orderly manner. Remember the key questions:

1. What have I got?

2. Where is it?

3. Where is it going?

4. What have I got to stop it?

5. What do I do?

6. Where can I get more help?

7. How am I doing?

8. Can I terminate the incident?

These are not terribly complex questions, but they can guide you on the path from blazing building to extinguished fire. What I did in this book is seek to pound them into you in a step-by-step fashion. I started with the basics of our business, then covered the need for sizeup and the like. Then I moved on to discuss those simple incidents you might encounter just about any day in your career. Next came increasingly difficult levels of fireground contact. Remember, no matter how easy the situation is or how difficult it becomes, your response will be the same. The questions remain the same. Just remember that the number of extra facts that must be assimilated will grow. But if you are familiar with the system, you will move right through it.

I will be most pleased if, after a particularly difficult fire, you pause a moment and reflect on what you have accomplished. And if in that moment it comes to you that something from this book made your job easier, I will be pleased as punch. I will never know about these incidents, but will still be happy. That is why I wrote this book: to make your job easier.

Thank you for taking the time to learn about **Fire Fighting Strategy and Tactics.** Throughout my career, I have been a people-oriented firefighter and fire officer. From my earliest experiences as an Air Force Fire Fighting Crew Chief during the Vietnam War, through my company officer years, right up to chief-level command, people have been my orientation.

One of my favorite descriptive phrases sums up my philosophy. "Nobody ever put out a fire with a white helmet and a portable radio. It is those fine folks with the black helmets and grimy faces who get the job done." Those are the people to whom this book is dedicated. I guess if you are a chief who is willing to learn, then maybe there is some help for you within the pages of this presentation.